MORE *Adventures of Fish & Game Wardens*
VOLUME 2

VERMONT WILD

MORE *Adventures of Fish & Game Wardens*
VOLUME 2

Written by **Megan Price**

Illustrated by **Norma Montaigne**

Pine Marten Press

Pine Marten Press

First edition

Copyright © 2012 by Megan Price

Book Design: Carrie Cook

Printed in the USA

For information and book orders, visit our website:
www.PineMartenPress.com

ISBN: 978-0-9828872-0-2

Library of Congress Control Number: 2010934765

Answers to your most pressing questions:

Did the stories in this book actually happen?
Yes. We couldn't make this stuff up.

Have the stories been embroidered just a little?
A whole lot less than most huntin' and fishin' stories.

What about the characters?
In most instances, we've used the real names of
wardens, deputies and innocent bystanders.
But only after they swore they wouldn't sue us.

What about the poachers?
We've changed their names and
sometimes fudged the location just a bit.
All the cases and convictions are real.

Get Ready..... Here it comes....

BIG DISCLAIMER
Any resemblance to any individual, living or dead,
is one heck of a coincidence.

That's our story and we're sticking to it.

Dedication

This book is dedicated to
Constance Jane Williams
who instilled in me a
love of reading and writing
and wild things, and to
Albert Evan Price
who took his daughter fishing
and hunting just like the boys
and
to all who work to protect
wildlife and wild places.

Ten Percent

Ten percent of all profits
will be used for wildlife protection
and outdoor education.

Look for Volume 3 of *Vermont Wild*
coming soon...

STORIES

"I saw the Chevy's front bumper was hung up on a boulder the size of a dorm refrigerator. My rear axle was buried in the mud."

4WD

Contributed by Eric Nuse

Four wheel drive trucks were introduced to Vermont game wardens in the 1980s.

Going the way of the dinosaur were our state issued Plymouth Furies. Those rigs were really built for the state police—low, lean, chasing machines designed to nab speeders and respond quickly to callouts and crashes over paved highways.

Wardens tried to make them work, but the big cars weren't very practical.

You'd find yourself in a chase trying to catch a poacher slamming through a rutted farm field or old woods road and the gas tank would get jammed and scraped across a hidden rock and punctured or the whole bottom of the cruiser would get hung up.

We were up against some enterprising poachers. Many had their own 4WD trucks.

Others created woods buggies or doodlebugs in their backyards for fun—welding left over car and truck parts together.

To give credit where credit is due, these guys were some of the creative geniuses whose tinkering contributed to the creation of today's popular sport utility and all terrain vehicles.

But with the introduction of 4WD Chevy 1500s and Dodge Power Wagons—suddenly, wardens could go where no warden had gone before—on four wheels anyway.

Of course, it took awhile before some of us learned the limitations of 4WD.

Mistakes were made.

The trucks we were assigned were a far cry from the cliff climbers folks can buy today. There were no fancy space age ultra grip tires costing $1,200 apiece or instant computer analysis regulating which wheel spins when in response to road conditions and all that fancy stuff.

We had posi-traction rear ends, two or so more inches of clearance and manually spun hubs that put us—at last!—in 4WD.

I was itchin' to see what the truck could do.

And that got me into some hot water.

I was patrolling the East Hill Road in Eden, a few days before deer season was to start when I heard two shots in quick succession and headed for the sound.

I rounded a bend and came upon an open top woods buggy with two fellows in it about 700 feet in front of me.

They'd heard my truck coming. But when they saw me come around the bend in the familiar green of a state truck, I think that shocked them.

I wasn't close enough to see their faces well. But I saw them look at me and then engage in a brief conversation.

While they chatted, I continued to roll up on them slowly and check out their vehicle.

Someone with metal shears and a talent for welding had had a lot of fun with this who knows what it once was.

The thing sat up in the air a good 20 inches off the dirt. There was a welded steel cage over the top of it in case of a roll over, a 2 x 12 piece of hardwood on the back for a rear bumper behind

some very gnarly oversized mud chewing tires.
No tailgate and no license plate that I could see.
Just some brake lights on the back.

I was looking forward to having a little chat.
But their conversation ended quickly and the
driver hit the gas.

That was my second clue they were up to no good.
So I took out a blue light with a magnetic base
I had on the seat beside me, turned it on, rolled
down my window and lobbed it up onto my roof.

If they wondered who I was before, well, now
they knew. I was the law.

I stepped on the gas and tried to catch up to them.

They went faster.

I gritted my teeth and sped up too. I figured
I had a good chance of catching them in my
new Chevy truck no matter where they went on
these back roads.

But in about a mile, we ran out of town
maintained road.

A sign said so. Warned drivers of the fact.

But they just kept on rolling.

We weren't exactly at warp speed—nothing like a chase on the interstate. They might have been going around 30 to 35 miles an hour tops.

If someone was coming the other way—at us—I didn't want to run them off the road.

It was more of an "I know you want me to stop and I'd rather not" chase.

The passenger turned his head to look over his shoulder at me every now and again to see if I was still behind them.

His right elbow was hooked over the rolled steel beam above his seat as he tried to stay with the big bouncy buggy his buddy was driving.

With a shorter, higher rig, the outlaws could navigate this road a lot faster than me.

As the road got rougher, I became envious of the fact they didn't have a roof over their heads. Mine kept slamming into the Chevy's headliner.

I was being bounced around in my seat like a tennis ball slammed by a pro at Wimbledon.

I managed to stay up with them but I wasn't getting any closer.

In another mile, the Class 4 road opened up onto a high hay field, sitting across from a long abandoned farmhouse whose roof had collapsed into the cellar hole many years earlier.

It was a very pretty spot I hadn't visited before.

I made a note to myself to return here someday with the family for a summer picnic—if I didn't break my neck first.

Apart from a couple mud holes, the farm field wasn't bad going and I gained a good 150 feet on them.

Having determined the dodgy duo was not impressed with my blue lights, I decided maybe my siren might get their attention.

I sounded it and the pair turned around briefly and stared at me in between bounces.

I was close enough now to see their faces. They were sporting big confident grins. The passenger even waved at me with his free hand.

I wondered if they knew something I didn't about what was ahead. This wasn't the look of a couple of fellows who planned on pulling over.

I followed them into the woods at the far side

of the hay field and saw I was now on a fresh logging trail—with deep skidder tracks and brownie mix thick mud in front of me.

I clenched my teeth, stopped and jumped out of the truck to lock the front hubs into 4WD.

I was determined not to let them get away.

Then I leaped back into the cab, shifted the Chevy into its low range and headed in after them.

I felt like I was Chuck Yeager—not exactly about to break the sound barrier, but definitely a test pilot of some sort.

I went about 100 yards and hit a sharp left turn with wide dark puddles in front of me and the nose of a couple big boulders too.

I gritted my teeth and steered the Chevy up onto the opposite bank—trying to slide around and through this hairpin turn like a snowmobiler bouncing off a snow bank.

A scratch or two from the saplings ripping across the front fenders and doors wouldn't hurt the paint much.

But if I caught the bank with a fender, or an

axle landed square on one of the boulders in the middle of this narrow track, I'd be all done.

It was a desperate move, but my only other choice was to let them get away and quit.

My wheels rolled up the bank, arced around and landed back on the trail with the impasse behind us.

I grinned. No way could my old Fury have done this. My Chevy 4WD was doing great.

Of course, there were more obstacles ahead. So many of them I started to feel like I was in bobsled run through molasses—someone who just couldn't wait for winter snow.

No one in their right mind would try to drive a vehicle through this mess of boulders, mud, broken trees and ledge.

But here I was.

The passenger in the doodlebug turned and looked back at me. I was just 250 feet off the buggy's bumper now.

The smile had left his face. He looked concerned. But still the driver didn't stop. I knew there was a ridge ahead.

But exactly where this trail went, I wasn't sure.

I continued to pound through mud holes, roll over rocks and slide past saplings that yanked on my side view mirrors threatening to tear them off.

All the time I'm sinking lower into a rock and mud funnel.

The loggers using the old woods roads—I use the term "road" loosely here—were driving skidders about two to three times taller than my shiny new Chevy.

Their machines sported tires 10 times bigger and chains to boot. And the clearance those rigs offered was a good two feet too.

But in the heat of the chase, you don't much ponder the little things.

I kept going, thinking the road would open up any second now and offer my Chevy better footing and more elbowroom.

Instead the trail kept getting darker and darker. The tree canopy was closing in over me.

I start to get a funny feeling, like maybe this isn't going to turn out so well—even with my new 4WD woods horse under me.

Trouble is, there's no place to turn around even if I want to call it quits.

Maybe the outlaws up in front of me will flatten a tire or get hung up?

Because looking out my side window, I can see I am getting deeper into doo doo.

It was like I had driven into the center hole of one of those funnel cake pans the wife uses to make a pretty Bundt cake.

There's a fat funnel end and a skinny one on that pan.

I had a bad feeling I'm headed towards the skinny end and about to get my grill stuck big time.

I look up from the trail for a split second and see the rocking rear end of the woods buggy taillights up ahead. The driver is alternating between the gas and brakes, climbing a steep winding ridge maybe 400 feet away.

My truck is crawling now, the 4WD doing that "urrrurrurrurrrr" sound as it rolls along like a tank.

4WD

I think about jumping out of the cab and just running after these guys. It'd be faster.

But I don't dare stop for fear the truck will sink into deeper mud and get hung up on a hidden rock or two, maybe even tear off the fuel line.

I round a little bend and there in front of me is a boulder the size of a college dorm refrigerator smack in the middle of the trail.

No way can I drive over it.

I grit my teeth, punch the gas and swing the steering wheel hard to scoot around it.

That's when my luck runs out.

My left rear wheel catches on a six inch stump stuck sideways in the bank I am trying to use as a sideways springboard—to get over and around that hunka rock that is just waiting to rip the bottom outta my truck.

The back end of the truck shudders.
My truck is thrown sideways back into
the mud brown soup.

I hear a pop, bang and a crunch, a grind and a groan—the groan is from me—as my 4WD chariot shudders to a stop.

The engine stalls.

I throw the truck into park, count to two, then hit the ignition.

The engine roars to life. I pull the shift lever into low low again and try to go go.

The wheels spin and mud flies from the tires, but I'm not moving.

"Come on! Come on!" I'm pleading with the truck and squeezing the steering wheel as hard as I can, like pinching the plastic will get her to jump out.

More mud flying.

I try reverse.

Even more impressive mud flinging—along with the distinct feeling I am sinking even deeper into the brown batter.

I try the old Vermonter "rock the truck" trick a half dozen times—until the smell of burning rubber—or is it the transmission?—permeates the cab.

The heat gauge on the dash is headed for the meltdown red zone. I am royally stuck.

Dug in deep in the Diggins, to be precise.

Done in.

I throw the Chevy into park and turn off the ignition and bow my head.

I've followed those scofflaws miles farther than I ever could have in the old Fury. But my new 4WD showroom sled is still no match for the rig these backwoods bandits have thrown together.

A shorter wheelbase, big knobby tires and three times the clearance has allowed them to float through mud and over boulders I just can't cross.

It's saved their bacon—for now.

But they gotta circle back home eventually.

I make plans to tour more of Eden's back roads in the weeks ahead. Their no name rig is unique. I'll recognize it again when I see it.

We'll have a little talk then.

But right here, right now, I have a bigger problem.

I've got to get the truck—and me—out of here.

I look to my left and see a mossy ledge and tree trunks staring me in the face less than a foot away.

I open my door. I get it about three inches off the latch when it hits something and doesn't want to budge further.

I put my shoulder into the door and push.
I gain another half inch.

Now, I have about four inches of daylight between the door frame and earthen bank.

I pride myself on a trim physique, but I'm no tube of toothpaste.

Hunh. There's no squeezing out this door.

I slide across the bench seat to the right and look through the glass.

A boulder the size of a Volkswagen is perched precariously, just above the cab. A freshet of water trickles from beneath—which is why the mud hole here is so deep.

I don't even dare roll down the window. I have the feeling if I sneeze hard this VW sized rock might just roll onto the top of the truck and squish it and me like a bug.

Oh boy. I scrunch down and slide back
into the middle of the bench seat and sigh.

This isn't going to be fun or pretty.

Time for an uncomfortable radio call, followed
by an ugly exit.

I pick up the receiver and call into Lamoille
County Dispatch.

I'm pretty impressed when I get a response.
I figure I'm calling from something just short
of the bottom of a 12 foot well.

I give the Dispatcher the bad news I am hung
up and will be walking out.

"Do you need assistance?" she asks.

"Well, the truck does," I answer.

"Will a tow truck do it?" she asks.

"Negative," I say.

"What will?" she asks incredulously.

"A log skidder or a team of big pulling horses,"
I say.

"10-4," she responds. I hear her stifling a laugh.

I stuff my gear under the bench seat, remove my keys from the ignition, grab my hat, coat and flashlight and stuff them inside my shirt and pants.

Then I slide on over to the driver's door, roll down the window and face the inevitable.

I've buried the truck so deep I am going to have to slither out like a prehistoric creature transitioning from swamp to shore.

I reach for a stout sapling, test it to make certain it will hold me, then twist my knee up until it is about touching my chin, stomp a heel out the window and into the soft bank, reach for another sapling with my other hand, breathe, pull and repeat.

I wriggle, crawl, stomp and clomp up a good 10 feet until I am finally standing on the ridge, looking back down at my very stuck truck.

You know how they talk about out of body experiences where you are looking down at yourself but you're floating above?

Well, I had a hint of that experience right there.

I knew if I had messed up this brand new truck beyond the point of a lengthy tow and maybe some simple repair like a new muffler, my career as a warden might be dead.

I saw the Chevy's front bumper was hung up on a boulder the size of a dorm refrigerator. My rear axle was buried in the mud.

It was a good four miles back to a road where I could hitch a ride.

I'll spare you the details of my walk out of the woods and hitching a ride.

For the next day or two I borrowed the wife's car to get around.

I was waiting for a call telling me I could pick up my truck, when instead I got a call telling me to meet with a supervisor. Mine was out of town on vacation. A veteran named Roy Hood was filling in.

I didn't know the man well and didn't know what to expect. I personally thought I was doing one heck of a good job.

But as the fellow who had to justify any costs above and beyond the Department's budget to the Commissioner and the Legislature, Roy had received the bill for my new Chevy's tow and repairs. Apparently, he had some concerns.

I entered the office and found him tapping a pencil on the desk mulling over a requisition form.

"Sit down, Nuse," he said firmly and launched right in.

"Warden Nuse, do you know the real difference between 2 wheel drive and 4 wheel drive?" he asked me.

I shook my head and said, "No, Sir."

"Well, let me tell you," he said, glaring at me.

"When you guys were driving two wheel drive vehicles, you got stuck a lot closer to the main road and it cost me a lot less to tow your vehicles out."

Roy sighed, stood up from behind the desk, took a deep breath, walked closer and stopped about three feet away, casting a big shadow over me.

He shook his head and said, "Here's the thing all you boys need to understand: It takes me two weeks to get a new truck into this office."

He paused and raised his eyebrows to see if I was still paying attention.

I looked up at him like a first grader in the principal's office. I froze like a baby bunny.

"But you young wardens?" he said kinda soft and shaking his head in pity.

I held my breath.

"I can replace you tomorrow," he said with a little smile in his voice. He seemed to like the idea.

Roy took a deep breath, turned and walked back behind the desk. He sat down heavily in the boss's chair, sighed again, reached over to another one of the many stacks of papers piled 18 inches high and without so much as looking at me again said, "Dismissed."

"Yes, sir," I said.

I jumped up out of the chair and bolted—making certain to close the door very quietly behind me.

Whew.

I learned all I ever really needed to know about 4WD from Roy.

"I throw myself forward like the high
school football star trying to get that
extra inch at the goal line, and
swing big with my net."

STOWE TURKEY

CONTRIBUTED BY ERIC NUSE

When wild turkeys were reintroduced to southern Vermont back in the 1970s, no one really expected them to take off like they did.

New York State was nice enough to give us some birds and we settled them into Rupert, Pawlet and Castleton down in Rutland County and crossed our fingers.

They flourished. Within a few years, wild turkeys were being seen regularly in Bennington County and north into Addison.

But way up in Lamoille County—where I was working in the 1980s—a wild turkey sighting was still unheard of.

So, when the first wild tom in more than a century poked his head up out of the tall grass along the border of Stowe and Morrisville, he created quite a stir.

It was like a big star's tour bus had pulled into the area.

The turkey wandered about pecking at the ground feeding, fluffing his feathers, and now and again broke into a strut.

And that strut—where turkeys fan out their tail feathers and become a kindergartener's Thanksgiving art project come to life—was what everyone wanted to see.

His fans went wild.

It was kind of nice to see so many people of all ages getting excited at seeing a wild turkey. It's a lot of work over many years for dozens of unsung biologists to restore turkeys and other species to their native habitat.

The problem was the spot this tom turkey had chosen for his stage—Tinker's Turn on Route 100.

Tinker's is a long flat stretch of busy highway —well, long and flat and busy by Vermont standards anyway—with a dip and a twist and a tight turn at one end, just to keep it interesting.

The bird liked the fact there was a grassy field

and some small elms and an old barn on one side of the road and on the other a corn field with a quarter acre or so of yellowed stalks still standing and offering him food, from the year before.

He crossed the road several times a day and he didn't much care what all was barreling down on him in terms of traffic.

Farm tractor or tractor trailer, it didn't seem to register.

It was Spring. Maybe this tom thought one of the cars would be driven by a pretty hen and she'd pull over and introduce herself.

And when this turkey ran or flew across the highway, he scared the heck outta a lot of people.

Turkeys are big. Their wingspans can top five feet. They make crows look like chickadees. And they're not the handsomest birds in the world either.

The sudden appearance of a bumpy blue and rubbery red waddle, inky black eye, snake like neck and Boeing 727 wingspan prompted many drivers to yank on the steering wheel, slam on their brakes and gasp.

This was very bad news for tailgaters.

Brakes squealed, bumpers got tapped, traffic came to an abrupt halt and police were called out.

And as word of the bird began to spread, people began showing up with binoculars and cameras and staking out the roadside to watch for him —hoping to see him do the famous Tom Turkey Strut.

So now the bird is causing traffic jams too.

The turkey is oblivious to all the havoc he's causing and while it's not his fault, the Stowe police have to deal with the mess.

Forty vehicles jammed up on either side of the narrow shoulders of Route 100 might not be much of a traffic mess when compared to Los Angeles, but it's a lot for this stretch of Route 100.

I get a call.

"Ya gotta do something about that turkey," was the gist of their message to me.

Well, wild turkeys were new to me too. I'd read about them but I hadn't dealt with them first hand.

I was however familiar with chickens. We raised them at home. They weren't very bright. A little grain spread in a corner up against a shed, some methodical foot shuffling and you could catch one pretty fast.

Chickens tend to run in circles—squawking, flapping and making a scene. They are lousy fliers. Just stick with it and you're bound to grab one eventually.

How much harder could it be to nab a wild turkey?

I told the folks in Stowe I'd get right on it. Then, I decided I had better take a look at the landscape and come up with a plan.

I drove over and noticed a ranch house on Tinker's Turn.

I pulled in and spoke with the owner, a woman who said she enjoyed seeing the bird but about had a heart attack every time she heard brakes locking up and tires squealing—which was a dozen times a day or more.

I asked if it would be all right to use her property to park my truck while I worked to catch the turkey and relocate him. She was happy to help.

I let her know I'd be back early in the morning, when the tom would most likely be feeding in the corn patch near her house and there'd be less traffic.

This bird might like an audience, but I don't.

I went home and put my turkey relocation kit together. Basically, it consisted of wiping the cobwebs off my biggest fishing net—not quite big enough to hold Namu the killer whale, but big enough to land a nice Nova Scotia salmon or—in this case—a basketball with feathers I estimated at 15 to 20 pounds.

The net itself was made of some sort of synthetic material and had an eight foot metal handle.

I just had to get in close for a good overhead swish at the bird.

I was back on the road to Stowe a little after 5 am feeling pretty confident this would be a piece of cake.

Stealth and intelligence should win the day over a birdbrain, right?

I parked my truck in the driveway, grabbed my fishing net and staked out a spot near a big old sugar maple in the yard. I keep an eye and ear on the corn patch.

I wait.

Hunters will tell you turkeys get up with the dawn, but if they do, some must spend a lot of time in the birdy bathroom or something.

Or at least this particular tom did.

He was in no huge hurry to get breakfast.

It was close to 9 am before I heard rustling off in the corn.

I get into a crouch and begin sneaking up on him with my net in hand. I figure he'll be so busy scratching at the corn he won't even care about me.

The homeowner said she had seen people get pretty close to him.

My experience with my hens and a couple testy roosters at home was similar. You throw them some feed and they could care less about humans walking around—until it's too late.

So, I figured the same casual stroll by technique would work here.

Wrong.

I get within 20 feet of him and he throws his head up, makes a terse "putt putt" sound and boot scoots deeper into the corn rows as slick as an otter slides down a river bank and disappears into deep water.

Gone.

Well, he's definitely a lot more wary than my chickens.

Hunh.

But I wasn't going to quit that easily. I decide to circle around, look for his tracks, listen for him and try this again.

It's past 11 am now.

The sun is getting hotter and a crowd of maybe a dozen people has gathered at the side of the road watching me watch for the bird.

Just what I didn't want. Oh well, all the more reason to get inside the corn patch. I can hide there.

I slip out of sight of the crowd, and try sneaking up on the turkey. I can just see his feet maybe 30 feet away, scratching.

Crouched low, unfolding one leg at a time as
smoothly as a leopard, no sudden moves, the
net out in front of me—I am ready to swoop.

Sweat is running into my eyes. The paper thin
corn leaves are tickling my nose.

He sees me coming and when I get within 15
feet of him he stops feeding, snaps his neck up,
tucks his wings at his side and beats feet away
from me again.

Dang.

No running in circles, cackling, flapping and
fussing alarm sounds outta him. He just makes
another beeline boogie away from me.

He reminds me of Jed being chased by Granny
in countless episodes of the Beverly Hillbillies.
And right about then, I'm behaving like Granny
—muttering under my breath and shaking
my head.

The only thing missing was Granny's cast iron
skillet. She was always sorta swinging that pan
around like she was dying to hit something with it.

I had my net—but it wasn't nearly the same.
Too light and I couldn't afford to bend the pole
anyhow.

I take a couple deep breaths and try again from
another angle.

After a couple more failed sneak and swoops,
I decide the problem is the pole.

It just isn't long enough.

So, I walk back to my truck and drive on over to
a building supply store.

I pick myself up an eight foot piece of iron pipe
and a roll of duct tape.

And from there, I head over to Lamoille Grain to
survey their turkey feed.

Once I explain to the folks behind the counter
what I'm up to, they donate a busted up bag of
cracked corn that had been torn in transit
to the cause.

It's late in the day now and I figure I should wait
until the following morning to try this again.

I don't want to mess with rush hour traffic and I
am hoping the bird will forget about me.

How long can it take for a bird with a head the
size of a dieter's entrée and a brain the size of a
dried pea to forget, right?

I spend the rest of the day on other duties, go to bed early and am headed back to Tinker's Turn by 4 am.

I move quickly to scatter a trail of grain from just inside the edge of the corn patch onto the lawn leading past a big maple tree.

My plan is for the bird to peck and move, peck and move and when he isn't looking, leap out from behind that maple, lunge and net him.

With the net handle stuffed inside the iron pipe and the two duct taped together, I now have a reach of more than 12 feet.

That should be more than enough, I figure.

By 5 am the trap is set and I'm ready—my pole standing tall against the tree trunk and me sitting quietly beside it, waiting.

Trouble is, once again the tom is keeping rock star hours. I don't know if he wears shades to keep the sun out of his eyes or he has other people to see and places to go before he heads over my way or what.

But by the time I hear him coming, I've been sitting with my behind on the cold ground and my legs out straight for more than four hours.

I can't feel my legs—other than the tingling from them going numb.

When I finally hear the tom scratching off to my right, I wriggle my toes inside my boots to try and get some feeling back in my feet.

I take a deep breath and look out of the corner of my eye. I see him sauntering out of the corn patch, onto the lawn and pecking at the corn trail, just like I'd planned.

I'd been a little stingy with the kernels in making the trail towards the tree. But I'd laid out a good heaping pile where I wanted him to end up—like gold at the end of the rainbow.

My plan is to jump up from behind the maple, swing big and net him while he's busy gorging himself.

My heart is racing as he comes in closer and closer. I see he's a mature bird, got a good three to four inch beard and a blue and red face only another turkey could love.

Within three minutes, he's hit his mark.

I keep my head still and reach with my left hand for my net handle.

I close my fist around the pipe, lift up a couple inches and pull the net towards me.

That's when the top of the net ticks a limb above me and a whisper of a breeze lifts the tail up a few inches like a windsock.

That's all it takes.

The tom's head shoots up and his neck goes straight as a shotgun barrel.

He pinpoints my location better than any multi billion dollar Pentagon purchase could ever dream of doing.

I'm found.

It's do or die time.

I lunge around the tree, leap forward and swing the net with both arms out straight.

Anyway, I try to.

The problem is my legs are numb.

I fall forward like someone had hit me with a shot of Novocain in each thigh an hour earlier.

My legs are silly putty.

The run and leap part of this turkey trap plan is not happening.

But I don't quit.

I wriggle ahead like the high school football star trying to get that extra inch at the goal line and swing big with my net.

Swooshhhhhhhh goes the net as my legs fold up like cooked spaghetti.

I face plant right into the lawn. But I keep my white knuckle two fisted grip on the pole.

I lay there—my nose deep in the wet earth like a robin's beak after a spring rain—hoping to feel some resistance inside the net telegraphing up the pole.

Maybe, just maybe the bird is inside and my plan worked.

One, two, three... Nothing.

I raise my head, spit out a mouthful of grass and see the net is empty and the turkey has skedaddled again.

I hear some murmurs behind me. I see some of these people are pointing and nodding

to one another. I guess they are all playing
Monday morning quarterback, critiquing my
technique.

There's a crowd of maybe a dozen people—
some with cameras around their necks, some
with coffee in their hands—some with both—
watching all this from the road shoulder maybe
75 feet behind me.

I'm grateful there's no loudspeaker, TV
chalkboard or instant replay.

I take a deep breath, climb to my knees and
attempt to walk.

My legs and feet are all pins and needles,
coming back to life, but not fully functional.

I stumble like a runway model that's lost a four
inch heel, back to the big maple, dragging the
net behind me.

I look over at my net and think maybe the
extension is part of the problem on this try.

I'm not sure what law of physics would apply,
but I know some professor could dazzle me into
a hypnotic state explaining why the arc of the
hypotenuse and the pi of the circle caused my
swing to be delayed blah blah blah.

All I know is that is one fast, lucky bird.

I shake my legs, one at a time, to speed some feeling back into them. And as soon as I can walk a straight line, I head over to talk to one of the police officers directing traffic.

He confirms what I suspect—the turkey is roosting in a clump of elms on the other side of Route 100, maybe 150 feet back from the highway where two neighbors' property lines conjoin.

Knowing where the bird roosts at night makes me think maybe I can catch the tom when he's asleep.

If nothing else, at least there won't be an audience to watch me.

I thank the officer and head for the elms. I find a half dozen of them all slowly succumbing to Dutch elm disease. There's a pile of bird droppings below a stout branch about 15 feet above the ground— looks to me like this is the tom's roost.

I circle back to my truck—avoiding much free advice from the crowd watching the turkey feed and strut—and get Jeff Hill of Morrisville on the radio. Jeff worked with me often as a deputy.

When Jeff hears what I'm up to, he's in.

Jeff tells Jackie, his wife—who is no slouch
when it comes to the outdoors either—about
this turkey trapping effort. Jackie's eager to
help too.

By nightfall, we add one more—Jackie's dad,
Robert Magoon, the county's probate judge and
an avid hunter and angler.

I ask Jeff to bring the gang and meet me at
9 pm at the ranch house. I advise them to
bring fishing nets, flashlights, headlamps or
what all they think might come in handy
and to be certain to wear long sleeves and
gloves.

I'd heard turkeys can really do some damage
with their spurs when up close and personal.

Shortly after 9 pm, we assemble in the driveway
of the farmhouse and come up with a plan that
involves luck and nets.

We walk across the road in single file, with me
leading the way, and nets at the ready.

There's not much of a moon, but I can see
the outline of the turkey on that elm branch
I located earlier in the day.

We all try to move in as silently as we can.

Maybe we looked like upright coyotes. Maybe birds never really sleep.

I'm 25 feet from the tree with my posse right behind me when I see the tom's head rise up slightly and his wings go wide.

He looks huge in the dark—like a small plane. But he doesn't bother contacting the tower for clearance.

He just launches.

Whether out of pure dumb luck or clever calculation, the bird heads for Jeff.

Jeff jumps and swings his pole, but the bird clears the net by two feet and keeps on gliding.

The bird has spotted us. There's no point being quiet.

"Get him!" I shout to inspire the troops and all four of us are off and running. The bird appears to be coasting towards a grassy landing maybe 50 feet away.

"Try circling him," Jeff suggests and so we split off into the dark, fanning out with a loose idea of pushing the bird back into one or the other of us.

Maybe then we can pounce and snag him in a net or under the blanket that Jackie is carrying. It's a good plan, but this bird is part magician. He astounds us with his ducking, hiding, scooting, sneaking and flying skill.

For a good five minutes we run around like crazy people in the dark with nets, flashlights and miners' beacons turned on bright, then turned off, nets raised, then flopped down.

We trot, trip, stumble and fall repeatedly in the dark field chasing this thing.

Then I hear the judge yell, "I think I got him!" and he trots parallel to our frequent flier's most recent run and attempts to cut the feathered fool off.

Good plan, but this isn't a pint sized partridge.

The big bird blows up out of the weeds just two feet from the judge's nose—all five feet of the big tom's wingspan flapping furiously into the judge's face.

The bird's spurs are out straight and pointed right at the judge's chest.

I see wings encircling Judge Magoon's
head and a kuh-fwahpping, kuh-fwahpping,
kuh-fwahpping sound as the bird boxes the
judge's ears once, twice, three times and digs
his big feet into the judge's coat.

The judge does the only thing he can—he
throws himself into reverse, slips on the dewy
grass and falls flat onto his back.

With the judge out of the way, the turkey
catches a breeze under his wings, rises up
and spins in the air.

Jackie is right behind her dad running
with a blanket in her hands, ready to pop it
open like an accordion and pounce on
the bird.

The problem is her father is on his back trying
to recover from the unwanted turkey hug.

Jackie is about to trip over and land right on
top of her dad.

Lucky for the judge, Jackie's nimble.

She jumps over the judge as nicely as any
NFL receiver and then dives forward in a
desperate attempt to cover the bird with the
blanket.

It's a valiant effort, but once again the bird is too fast.

Jackie lands spread eagle in the grass a few feet from her dad.

The turkey sails off into the night once again.

Jeff and I run over to make sure Jackie and the judge aren't hurt. We find the two of them laughing hysterically and rolling around on the ground.

Jeff and I switch on our flashlights and shine the field.

There's no sign of the bird.

The four of us traipse around the field another 30 minutes or so, but it's hopeless.

The bird has been lost to the landscape once again.

It's close to 11 pm and we haven't so much as a tail feather to show for our efforts—although the judge claims he may have swallowed a couple fluffy little turkey breast feathers when he and the tom bumped chests.

I laugh at his joke. But I'm pretty frustrated.

I'm not looking forward to more phone calls in the morning as the traffic jams and fender benders continue, and the police and public call 1-800-ERIC.

I'm one step away from using lethal force on this feathered friend.

The four of us walk back across Route 100 and head home. It's close to midnight when I finally walk into my house.

The family's asleep but I'm still stewing about this bird. I peruse my hunting books, searching for some new ideas on how to capture a wild turkey.

And that's when I hit on it—just a couple paragraphs claiming some resourceful settlers used whisky soaked corn to get a turkey.

It wasn't real sporting, but if you're starving to death and hunting with a musket that can barely hit the broadside of a barn at 20 paces, you gotta try and give yourself some sorta advantage, I guess.

The book claimed the whisky mash slows the bird's reflexes enough so a pioneer could get off a shot that might actually knock the bird down.

Well, it's all I can do not to start whooping and

hollering right there in the living room. But I
didn't want to wake the whole house up.

Instead, I jump up and tear into my stash
of booze—a dusty collection of cheap caramel
colored poison. Some of it has been sitting
so long it's the consistency of 50 weight
motor oil.

I line up a half dozen bottles of stuff I know I
will never drink, dust off the labels, unscrew
each cap and take a sniff—one at a time.

At least two smell like paint remover. I can't
recall if turkeys have a good sense of smell, but
I decide not to chance it.

I pick out a bottle whose contents smell the
least disgusting but who's label promised a big
alcohol punch—a brew called Rebel Yell.

I know there are healthy turkey populations in
the southern states. Maybe this bottle is a sign
of some kind.

I grab a bucket from the mop closet, run out to
my truck, pour the last of the donated corn into
the bucket and empty the entire bottle of happy
juice on top and kinda slosh it all around.

I decide to just let that corn soak overnight.

Come morning, I'll see if I can get this bird drunk. Maybe then I'll have a fighting chance of catching him.

If this doesn't work, I'm going with the cannon.

I went to bed and set my alarm clock for 4 am.

By 4:30 am, I am back out at Tinker's Turn, with my long net pole still intact and my new secret weapon sitting beside me in the front seat —whisky soaked cracked corn.

After two days of chasing him, I knew where he traveled and where he grubbed for corn and insects. I had no idea how much boozy feed he could or would eat or how much it would take to get the tom drunk, but I wasn't going to hold back here.

This was do or die day for the bird. He was coming with me today one way or another. I couldn't risk getting people hurt even if he was the first turkey to be seen in Lamoille County in more than 100 years.

I set piles of whisky soaked corn all along his route on both sides of the highway.

If I made a few crows or raccoons tipsy in the coming week, so be it.

After I'd set out all the doctored grain, I decided it was time to wait him out and give myself a little treat.

I jumped back into my truck and drove into Stowe to grab a nice tall cup of black coffee and the local newspaper.

I was back into the lady's driveway by 6 am. I grabbed my net and my binoculars and trotted across Route 100 to see if the bird was off the roost yet. The sun was up enough for me to sneak a peek from a distance.

He was off the roost—although I couldn't be certain he'd even returned there after our nutty rugby match of the previous night.

How long does it take a turkey to get drunk on liquor laced corn? The book hadn't said.

I decided to give the bird an hour to tuck into his breakfast and eat himself into a stupor before I went looking for him.

I went back to my truck and glanced at the headlines in the Times Argus and made certain I wasn't in the obituaries while I enjoyed my coffee.

Every few seconds I pop my head up looking

over at the corn patch just in case my suspect shows up there.

At the hour mark, I set out.

My net is over my shoulder and my revolver is on my hip.

I start with the feed trail I set out for him within a couple hundred feet of his roost.

I see something has been kicking the corn around and some of it looks like it might have been eaten. Whether it was the tom or some other critter that had found it, I couldn't really tell.

I decide to just keep walking slowly, check out his usual haunts as best I can, and see if I hear or see him.

I walk about 250 feet when I hear something rustle off to my left about 50 feet away.

I see a tangle of wild apple trees mixed in with some grapevine, one and two inch hardwood stems, goldenrod and other brush.

Could just be a squirrel over there running through leaves and looking for last winter's buried treasure, but it's worth taking a look.

I walk over as quietly as I can and see a big flash of brown make a zig zag dash beneath a twisted wild apple tree.

It's the turkey.

He's decidedly wobblier.

This was hopeful. But had he slowed enough for me to catch him?

And how am I going to get into that tangle with my net?

I decide I can't wait for him to come out. He might just slip off again. I have to go in after him.

I bend low to slide beneath the gnarled tree branches, brush last year's goldenrod heads out of my eyes and plunge in.

The tom hooks another left. I speed up, but my pole doesn't. The handle is so long I can't bend around the saplings and woody stems.

I'm like a pole vaulter in a cluttered antique shop.

I wrench hard on the handle, shake the pole free and forge another few feet ahead.

Now my net gets tangled in some honeysuckle.

I see the grass moving off to my left.

The tom isn't his usual rocket self, but he doesn't have to be when I am all wrapped up in this brush.

I'm about to lose him.

I run my hands up the pole and claw at the duct tape to remove my extension.

I'm ripping, stripping and using my teeth to separate the iron pipe from the original net handle. I can't move in on the bird until I do.

I'm keeping one eye on the turkey and am relieved to see he's having his own technical difficulties.

The grass is waving about 20 feet on the other side of the apple tangle but he appears unable to launch into turbo turkey and beeline away.

He's either bumped his head, twisted a drumstick or boozed up on Johnny cracked Rebel Yell soaked corn and he don't care.

But I still worry about what he's got left.

If he manages to spread his wings and catch a breeze, I could still loose him.

With a trail of duct tape stuck to my boot sole and flowing behind me like toilet paper, I throw the iron pipe aside, lower my net like a lance and go charging through the brush after the bird.

I bounce off saplings, get hung up on brush and trip over remnants of an old stonewall and barbed wire.

I'm a warden pinball—two steps forward, net caught, yank it free, another step forward, caught again—over and over.

Dang it.

I sure hope the booze is sedating the bird because this is the slowest chase ever.

When I finally bust out of the thicket and search the field grass, I find the turkey sitting in a heap—wings out like an airplane that's been tipped in a big wind.

His head is low and his eyes are glazed.

He tries to stand and run, but it's not happening.

He swivels and collapses in confusion.

He's mine at last.

But I don't take time to gloat. You never know what wild critters can pull off at the last minute.

I lower the net over his head to hold him there, then drop the handle and put a boot on it—just in case the bird somehow roars to life.

Then I put my leather gloves on. No way do I want to get kicked by his spurs should he come to life on the hike back to the truck.

I lift the net just enough to push the tom deeper inside, roll and twist the handle until the bird tumbles quietly into the pocket, twist it another half turn to seal the top over him, then heft the pole onto my shoulder.

Feels like the bird weighs a good 17 pounds anyway. All the corn I have been feeding him the last couple days probably helped plump him up.

He doesn't complain much. Just a little wing flap. No cackling protests or big flap like my noisy chickens.

This tom behaves more like a big guy who is all

done in after a big night of carousing—he just wants to sleep it off.

Well, he was up most of the previous night. Maybe that's why the booze is affecting him so strongly this morning.

I scoot over to the edge of Route 100, and prepare to cross.

There's no one around.

Here I am with the Star of Stowe over my shoulder and there are no witnesses?

Where's the marching band, key to the city and some reporters sticking cameras in my face?

None of that is happening. Such is the life of a game warden.

I cross the road and lower the net from my shoulder into the truck bed. I lean in and take a moment to study him.

He blinks an eye at me and I know he's still alive.

I adjust the net a bit to make him more comfortable, then jump into the cab.

I want to let him go as quickly as I can.

I drive over towards Mud City and the Morristown Corners area.

Near the base of Bull Moose Mountain, there's the Rooney farm. It's about 10 miles from Tinker's Turn as a crow flies, 15 miles by road.

I pull into a hay field road, shut off the truck and take the netted tom for a short walk.

I lay the bird gently in the grass and unfurl the top of the net slowly.

I'm careful to keep his big feet pointing away from me.

No way do I want to give him a reason or the opportunity to slam those big spurred clod hoppers into my chest like he did Judge Magoon just a few hours ago.

The judge was smart enough to wear a heavy vest. I've only got a khaki shirt on.

The tom struggles a little.

His eyes look brighter. He seems better coordinated than he was just an hour ago.

His boozy breakfast is wearing off.

I lift the handle and in five seconds, he's on his feet, out of the net and running—not perfectly straight, but a lot straighter.

All I can do is hope he makes it. There's lots of food for him to eat, good trees to use as a roost and water nearby.

I stop down at the main farmhouse and let the landowners know the turkey has been set free and ask their cooperation in keeping his presence under their hat.

I also ask them to give me a call if they see him.

I'd like to know if he makes it.

When I get back in my truck, I call the Stowe police to let them know their feathered traffic hazard has been relocated.

There's a mix of relief combined with a "What took you so long???" in the polite thank you on the other end of the line.

I head on home to catch a couple hours of shuteye.

About two weeks later, I get a call from my friends at the farm.

"Eric—you know that turkey you dropped off?"
Joe says.

"Yeah. How's he doing? You see him much?" I ask.

"Yes, he's in the yard a few mornings every week
—mixing in with our chickens, picking at their
grain.

The thing is…"

"Oh no," I think to myself. "Here it comes.
This tom is causing trouble out on their farm
too!! I don't want to have move him again.

Puh-leease!" And I wince.

"… there's another turkey with him," Joe says.

"Really?" I grin into the phone.

"Yeah, it's smaller and there's no beard.
That's a hen turkey, right?"

"Sure sounds like it," I say and chuckle into
the phone.

And it was.

Now, I can't say with any certainty that this

pioneering pair is responsible for many of the turkeys we enjoy seeing around Lamoille County today.

But after all Judge Magoon and Jeff and Jackie and I went through to catch that wily bird—I sure like to believe it's true.

"And there it was—plain as day—a thumb wide trail of glowing bluish light running from the back of the head of this fisher a good three inches or so down the middle of its shoulders."

TICK TRAIL

CONTRIBUTED BY RICHARD HISLOP

Over near Montgomery, there was a guy named Clayton who was just too good to be true.

I don't mean he was too kind or too brave.

I mean fellow deer hunters and trappers figured Clayton just couldn't be as good at hunting and trapping as he claimed to be.

No one could.

Opening day of deer season, Clayton invariably showed up at one of the weigh in stations with a big buck and a story to match.

Of course, even the very best hunters fail to put venison on the table some years—let alone show up with a trophy every time.

And as if repeated success in the local buck pool wasn't enough, Clayton was also the king of trap lines.

Whatever was in season—muskrat, beaver, bobcat—he'd bring in more pelts than anyone else.

Lewis and Clark would have been impressed.

Pretty unbelievable.

Clayton's good luck raised a lot of eyebrows and started some whispers. While some of the grumbling was likely sparked by jealousy, others were genuinely concerned this guy was maybe doing some serious damage to local wildlife populations.

Clayton liked to run his trap line in the Witch Cat Falls area of Bakersfield. Other trappers would run across Clayton's traps and—like guys comparing car body work or carpentry—check out his sets.

A few of them had shared their theories with me as to how they thought Clayton was cheating.

Problem was, surveillance takes a lot of time and resources when there's a hundred other places to be and jobs to be done.

I knew I would need a lot of help to get this job done.

I put in a call to fellow warden, Eric Nuse, over

in Johnson. Eric had heard the same rumors.

Together, we hatched a plan.

Trappers had only one week to set their traps
for fisher. Because fisher pelts were selling for
very good money, we knew Clayton would be
after them.

Eric and I agreed that translated into a nice
opportunity.

For those that don't know, fishers are members
of the weasel family, dark brown or black in
color. They can be up to four feet long, with
the average adult weighing between eight to 12
pounds.

Because fishers are elusive and smart they
are not easy to trap. But if you could get 'em,
the money was good.

Most trappers were lucky to report three or four
fisher at the end of the brief trapping season.
Clayton always reported nearly twice that
number.

But Clayton was not a fellow you wanted to
confront without solid evidence in hand. He was
a big gruff grizzly type with enough hair on his
face and sprouting out of the top two buttons of

his camo shirt to possibly qualify him as a new fur bearing species.

He packed a pistol on his hip, a long thin skinning knife on his belt and a big chip on his shoulder in all seasons. He considered himself an expert in all things outdoors. Didn't matter if it was pickerel fishing or bear hunting, he'd done it, and he'd done it better than most anyone around—in his mind anyway.

Now, for fishers, the law required that each animal taken had to be inspected and tagged by a warden within 48 hours of the close of the season. Without the tag, a trapper could not sell the pelt.

Each warden carried sealing pliers with our identification on them. After inspecting a pelt, a warden stamps the wire into a soft metal seal so that our identification shows up. We also record the name of the trapper, his license information and the number of fisher he presented to us, along with the date and the town in which the animals were taken.

That data goes back to Waterbury where biologists analyze the data to determine the health of the species. Trappers provide a valuable service, just like hunters do. Biologists can't be everywhere and the animals hunters and trappers report help create management plans for many species.

Clayton hated this inspection process.

He would make his resentment very clear every year.

Typically, I would come home to a message on my answering machine on the closing day of fisher season. Something like, "Clayton here. Got fishers for you to tag," and that's all.

When I showed up at his home, his anger towards any and all government regulation would rear its head.

"Bunch of nonsense you coming on out here," he'd say and spit into the late fall leaves and snow covering what was once a lawn.

"A man oughta be able to hunt and trap and take whatever he needs and be left alone."

I would have preferred he kept his opinions to himself. But I just let it go in one ear and out the other.

If Clayton were a true student of history, he'd have known that for most of those good old days, his fisher traps would have been empty, precisely because there was no regulation and reporting.

While the wanton massacre of millions of buffalo and the extermination of many more millions of passenger pigeons gets all the attention in the history books there are plenty of other species that came close to the brink— even in Vermont.

Among the near casualties in the Green Mountains were moose and beaver. Both were plentiful when the Abenaki and other native peoples populated the northern forests. But as soon as the Europeans began settling here in the late 1700s, things changed.

The beaver were killed both for their pelts and because the settlers saw them as pests who flooded their farmlands. Beaver dams were blown up, wetlands drained. Moose, who benefitted from the beaver marshes vegetation for their preferred diet of twigs, were shot and eaten.

Both were about wiped out due to a lack of hunting and trapping regulation.

Even the native wild turkey was lost to Vermont and wasn't reintroduced until the 1970s—all from folks simply taking too many.

Fishers—which Clayton loved to trap because their thick fur brought him high dollars—were among the species Vermont had all but lost.

In 1929, the Vermont Legislature closed the fisher season, putting them off limits to everyone in all seasons. There were only a few left in the state.

What saved fishers—and got legislators, camp owners, trappers and woodsmen's attention —was fishers' fondness for dining on porcupine.

Fishers are one of a very few species that has figured out how to get past porcupines' protective quills—a trick most dogs never learn.

In the 1950s in Vermont, porcupine populations were out of control, girdling and killing thousands of trees with their sharp teeth—like some sort of mountain beaver.

Hard to imagine now, but there were multiple reports of porcupines chewing isolated log cabins and sheds along the Long Trail into sawdust.

The Legislature placed bounties on the porcupines to try and control their numbers. In one year alone, more than $160,000 was paid out to porcupine hunters.

Those efforts hardly made a dent. Porcupines kept up their toothy attack.

Biologists determined the solution was to bring back the fishers—pesky porcupines' natural predator.

In 1958, Vermont's Forest and Parks Department proposed to reestablish fishers in the state to restore balance.

Over a period of eight years, the Vermont Fish and Wildlife Department in cooperation with the Unites States Fish and Wildlife Service, live trapped 124 fishers in Maine, then drove them to Vermont and released them in more than three dozen communities around the state.

It took awhile, but by 1975, the fishers' reintroduction was considered complete.

What a bounty could not accomplish, the reintroduction and protection of fishers did. Porcupine populations were brought under control thanks to the fishers.

And these often maligned animals have been doing a good job ever since—unless of course, you see the world from a porcupine's perspective.

If Clayton were the kind of guy who would listen

and learn, I would have taken five minutes to explain this to him. But I knew I'd be wasting my breath.

Now, Clayton drove a pretty distinctive rattletrap with bondo and duct taped fenders. He was even nice enough to have the driver and passenger doors different colors. It wasn't difficult to identify his distinctive truck.

From past years' reporting, we also knew where Clayton liked to set his traps.

Add to this, the fact that fisher trapping season is in December, so we had some snow cover on the ground.

Eric and I kicked it around and came up with a plan. Eric would check Clayton's traps during the one week fisher season because it was a shorter drive for him to get to Bakersfield from his home. I would take over some of Eric's duties to help him out as an exchange.

Of course, one of the bigger challenges was to keep out of sight.

A warden has every right to check traps, but we didn't want to spook the guy. We wanted Clayton to stay with his routine and think he was alone out there.

With Eric checking Clayton's traps daily and sharing his information with me, we found Clayton wasn't visiting them every 24 hours as required by law. Eric also found illegal sets and traps without any identification on them—all of these are violations.

Still, we didn't want to rush in and flip open the citation book and start writing Clayton tickets. We wanted to let Clayton play the season out.

Now, this was before critter cams and the cool tech tools you see on television today. Patience, good legs, binoculars and brainpower are what we had at our disposal.

Which brings me to the use of deputies—unsung heroes in many enforcement efforts.

Deputies generally worked a full time job in addition to helping us out. We often asked these fellows to help us with long stake outs at night sitting in the woods, to give up time with their wives and kids—all in the hopes of maybe making a case against some poacher.

John Cushing of Milton was among my best. Whether sitting at the edge of a meadow at 3 am in late October waiting for poachers to shine a light or slogging through icy water checking goose hunters' stamps, he was dedicated.

By day John served as town clerk and treasurer for the town of Milton. But he's a pretty handy fellow outside of the office too. John was a great help to me on many occasions.

So, we invited John to join us on this case and along with Eric, the three of us came up with a plan to trap Clayton.

There was a new dye that had just come out—one that could only be seen under black light. The ink was in a base of something like Vaseline. It was slippery and tough to wash off.

It would be hard for Clayton to spot and wouldn't harm the quality of the fur at all.

Eric took the bottle of dye, a small paintbrush and off to the woods he went to trek Clayton's trap line.

Every time Eric located a fisher in Clayton's traps—or in a trap without proper identification which he believed to be one of Clayton's—the critter got a good spot of dye painted on the back of its neck.

The fishers didn't put up a fight—they had been killed instantly when they stepped into the conibear trap.

It was a long week of checking traps, avoiding Clayton, and gathering evidence.

When the week was over, Eric had marked four fishers.

If a trap had a fisher in it, but the trap was legally set, Eric did not dab it with dye.

Now, it was time to see how many hides Clayton would report and spring our trap.

Sure enough, Clayton called me the day fisher season closed and asked if I would come and inspect and tag his animals.

He didn't say how many animals he'd taken and I didn't ask.

I didn't want to appear too curious.

Given Clayton's bear like stature, his tough guy anti government survivalist talk and the fact I would be entering his den—literally—I knew this could kinda go bad real fast if not handled with a diplomat's skill.

I wasn't afraid of the guy.

But there was no reason risking him or me getting hurt. It's just smart to bring backup.

But of course, I needed a reason for other fellows to be coming along with me on a simple tagging job.

It wouldn't take much to make him suspicious.

Eric and I knew we couldn't just go marching into his house and down into his basement and start shining the black light over his pelts telling him we suspected him of numerous violations.

He could very easily blow up.

If that happened, Clayton would have the upper hand in terms of where weapons might be stashed.

We didn't want anyone hurt.

We needed a plausible story to get us inside Clayton's house.

Well, Eric and John and I got together and after a lot of brainstorming it was John who came up with the best idea—a really clever one.

John suggested he pose as a University of Vermont graduate student studying ticks.

"Ticks?" Both Eric and I raised our eyebrows at that suggestion.

"If there were ticks on the animals, wouldn't Clayton have found 'em by now?" I said, "or wouldn't they would have just dropped off after the animal died?"

"Yes, he's too smart to fall for this one, I think," Eric added.

John just smiled. He was a couple steps ahead of us already.

"Hold on. Hold on." he said, "There's more to it. I am using the black light to look for tick trails— to see where the ticks have been. It's new. And you can only see the tick trails with this special light," John said looking us each in the eye with deep sincerity as he spun his lie.

I stood there admiring his talent for creating a good yarn out of thin air.

"We need to check these animals for tick trails to see how the fisher population is being affected."

John was on a roll now with his story. He kept yanking line off his reel going farther out into ever deeper and uncharted water.

"The ticks could affect the fishers' health, including their ability to reproduce. If that happens, there will be fewer fisher to trap in the future," John said—still with a poker face.

By the time John got done with his tall tale about the ticks and the use of the black light to see imaginary tick trails—well, he just about had me believing it.

Eric and I looked at each other in amazement.

I shook my head and laughed.

John looked so sincere.

I believe he could have gotten a government grant if he'd just taken the time to write it all down.

One bonus we had going for us was John's youthful appearance.

This, combined with his ability to tell this tick trail tale artfully with a straight face made John's story our choice for approaching Clayton with a trap of our own.

Of course, a good story alone wasn't going to do it. We had to be able to get in and out of the house safely too.

Since Clayton had called me, I would take the

lead. Clayton also knew Eric. Eric had checked his pelts in other years, covering for me if I was tied up.

The good news was Clayton had never met John.

To justify the three of us paying Clayton a home visit, we came up with the idea that Eric and I were being trained to use this new tick tracking tool by John because the state was getting more concerned about tick damage to wildlife.

We had no idea if Clayton would buy the story —but it was all we had.

We'd all ride together in my cruiser. That left us some room in the back seat for Clayton if things got ugly and we had to bring him in.

I also knew from experience that if the three of us showed up in separate vehicles in his driveway, it could really trigger a Clayton rant about the "abuse of guvmint and taxpayer money."

The last thing we wanted was to get him riled up.

And three of us riding together made our story about visiting other trappers that day for the same research project, more plausible.

With our plan and John's story as good as it was gonna get, I gave Clayton a call and set a time to stop by.

I didn't mention our research project or that Eric and John would be with me. I didn't want to give Clayton a chance to say no.

When we pulled up the next day, Clayton was out in his driveway waiting for me and wearing his usual scowl.

I climbed out of the car and then Eric and John popped their car doors.

Before I could even utter a greeting, Clayton said, "What's this? You got company?

Since when does it take three men to check seven hides?"

Before he got his boxers in a bunch we couldn't untie, I smiled and said, "We're here on a research project, Clayton, and I thought you'd want to help."

I'd heard Clayton liked to think of himself as well read. And if reading Soldier of Fortune cover to cover every month was your idea of keeping up with current events, he'd surely qualify.

"Research?" he said. It was time for John to
step up and he didn't hesitate.

John walked on up to Clayton, reached out
a hand and introduced himself as a UVM
graduate student working on the effects of
ticks on various fur bearing mammals native to
Vermont—with fishers of particular interest due
to the economic boost they provide to serious
trappers like Clayton.

"Complimenting Clayton is smart, real smart,"
I said to myself.

"John's got his attention now."

I looked over at Clayton—he was listening hard
to what John had to say.

"The concern is that if ticks are doing damage
to these animals it could reduce the quality of
their furs. And if the ticks are really bad the
fishers may not be able to reproduce at all,"
John said.

I was standing back watching Clayton's face.
He was hanging on every word.

"Of course, if the fisher numbers go down, that
would negatively impact trappers in the state,"
John said.

Clayton looked like he could use a bit more information.

John got it.

"No fishers to trap, no pelts, no money for trappers like you," John summarized.

Clayton's head snapped back like a bee had just stung him in the ear. The possibility he might lose money had him on board.

"Hunh! That'd be bad, bad for me and for a lot of us trappers," he said. "But I've never found ticks on any of my hides."

"Right," John nodded and agreed with Clayton. "Lots of times you won't find a tick."

"But sometimes they've been there and damage was done," he added.

I saw Clayton's eyebrows shoot up.

"So, what do you use to find these ticks then?" he asked John.

"I use a special light—a black light they call it. I shine it a few inches above the animal and I look for a trail the tick leaves behind," John said with a straight face.

"Hunh," Clayton said again shaking his head as this news sunk in. "Ain't that something. I never heard of it. "

He looked at John like a kid in 6th grade science class eager to help the teacher put together a bottle rocket launch.

"And you got this light with you?" he asked John.

John nodded. "Yes, it's in the car. If you don't mind I'd like to use it to check your animals."

"Well, sure!" Clayton says. "This sounds like pretty important work. Though I don't think you are going to find any of those trails on my animals."

"Who knew Clayton was interested in science?" I thought to myself. "Always a mistake to sell someone short."

Eric and I stood back and just let John go with it.

He had Clayton eating out of his hand.

"Oh!" Clayton said. "Well, the skins are where I always keep 'em—in my den. You go on and get your light and we'll have a look, I guess."

Clayton took about five steps, then stopped and said, "Wait a minute."

The three of us stopped dead wondering if he was on to us.

"This light of yours, it doesn't hurt the skins any does it?" he said to John.

John stopped just short of the cruiser door, turned to face Clayton and said, "What do you mean?"

"I mean make 'em worth less when I go to sell 'em to my buyer," he said.

"Oh no," John said. "It's just a light shining. There's no heat or anything like that," he opened the cruiser door, reached in, and pulled the black light from the back seat.

Clayton nodded, turned back towards the house and resumed walking.

I looked at Eric and exhaled big. He nodded slightly in acknowledgment.

Clayton led us through the open cellar hatchway door, down into a small mudroom cluttered with boots for spring, summer and winter along with a series of traps and coats hanging on the walls.

The mudroom opened into a finished basement of his home. It was an original man cave.

When Clayton hit the light switch, the room came alive with three walls full of hunting, fishing and trapping trophies: a 10 point buck and four smaller ones—none less than a six pointer, two nice turkey fans with beards nearly six inches long, a big Muskie, brown trout and a host of other species tacked to his walls.

Impressive.

In one corner he had a wooden table with a revolver torn apart ready for cleaning or repair. Adjoining that table was a wall, maybe 26 feet long, lined with gun cabinets of various designs and vintages. Their glass fronts disclosed dozens of shotguns and rifles. Next to them were two deep gun safes sporting combination locks—where I assumed Clayton kept the real artillery.

"If Canada ever attacks, Clayton's ready," I thought to myself.

Clayton led John, Eric and I over to two eight foot long folding tables where the fisher pelts were lying side by side awaiting our inspection.

As I looked at the tables, I counted seven fisher

pelts. Once again, Clayton had about doubled what any other trapper would report.

Was it luck, exceptional skill or was he cheating? If the black light picked up any dye, we'd know at last.

"How's this?" he said turning to find John and adding, "Do you need more light or somethin'?"

"No, no," John said. "It helps if it's kind of dark actually. In fact, let me get away from these overhead lights. I'll get down here on the floor where it's a little darker. Then I can take a better look."

Another smart move by John.

With Clayton bending over John's shoulder and Eric and I standing behind the two of them, if Clayton got riled, we had a much better chance of controlling him.

Eric and I stood back and worked to keep a straight face as John got on his hands and knees and turned on his black light.

We had told John where to focus the light on the fishers to look for the oily dye. He would find it on marked pelts right at the base of their necks.

None of us had any idea which of these pelts John was inspecting would show the dye—or if any of them would.

For all we knew, none of these animals were ones Eric had marked and left in the traps.

John got to work.

"Okay, let me look," John said, and he bent down and turned on the light.

"Hmmmmm," he said, running the light one at a time up and down and over and over the first thick black fur.

John was like a mother looking at a skinned knee on a teary toddler who had just fallen off the backyard swing.

He spent a good two minutes on the critter, going nose to tip of tail on it until he shook his head, leaned back a little and said, "Nope. This one's clean."

There was double meaning in that verdict, of course, though it was lost on Clayton right then.

"Clean" meant free of ticks to him.

To me, it meant a term under the law called

"clean hands." So far, we couldn't prove Clayton had done anything illegal. This fisher had not been tagged with invisible ink.

I would tag it and Clayton could sell it.

John moved on to another skin and carefully went over the hide—same verdict.

I looked at Eric but didn't say a word. I'm thinking, "Are we wrong about this guy?"

Clayton straightened his back and stood tall and said, "Yeah, I really don't think you're gonna find any—"

When John—who like any good scientist was so intent on his work he wasn't really listening to us—said, "Ah! Here we go. I think I have a tick trail here."

Well, all three of us bent down to look.

And there it was—plain as day—a thumb wide trail of glowing bluish light running from the back of the head of this fisher a good three inches or so down the middle of its shoulders.

"Oh! Is that what they look like?" Clayton asked John, like a kid looking through a microscope for the first time at an amoeba.

"Yes," John said nodding and keeping a dead serious look on his face. "That's what we're looking for all right."

Again, John hit the nail on the head with his choice of words.

I had to lean back and bite the inside of my lip to keep from breaking out laughing.

I didn't dare look at Eric for fear he would bust out in a grin too.

"Well, I never," Clayton said, shaking his head and smiling a bit at the wonder of this new technology.

"Let's put this one back up on the table so we can take a closer look at it later," John suggested to Clayton. "Do you mind?"

"Oh sure, sure," Clayton said. And he picked up the pelt and laid it up on the table for us —literally helping us stack the evidence against him.

"Thank you," John said, and then he resumed his work, inching his knees over sideways across the carpeted floor to move on to the next valuable fur.

Clayton couldn't stand it any more—maybe his back hurt or maybe it was pure scientific curiosity getting the best of him.

"Hey, uh—John, is it?"

John sat back a bit and looked up at Clayton.

"Yes?" he said.

"Uh, do you mind if I give it a try? With the light, I mean?"

"Oh sure," John said, and he threw his knee out big and scooted over about three feet to make room for colossal Clayton to settle down next to him.

Clayton got down on his knees like a horse about to take a nap in his stall. It wasn't pretty. There was a lot to the man.

Once on the floor, he bent over at the waist, got his shoulders hunched forward, elbows bent low and waited.

John handed Clayton the light.

I can't believe my eyes. Clayton is searching for tick trails alongside John.

"Is this how you do it?" Clayton asked John.

"Yes, you just go slowly over the hides—just like that—and look for the bluish reflection," John said.

John kept careful oversight of his student's work.

"Oh! There's one!" Clayton said.

"Where?" John asked.

"There! On the back of that fisher. Is that one? A tick trail? Am I right?" Clayton asked.

"Yes, you found one," John said.

Clayton smiled big. He was having such a good time playing scientist alongside John.

I leaned back and looked Clayton over more carefully.

I was looking for any evidence of a knife in his boot or the outline of a gun in his pocket—anything that might be used as a weapon against us that would be within his reach.

There were the usual table lamps and chairs in the room.

And, of course, it wouldn't be a lot of fun to get smacked upside the head with a fisher pelt either.

I was hoping we could give him the bad news in a manner that wouldn't rile him up too bad.

We were going to have to spring our own trap soon.

But Eric and John and I had a exit plan there too.

In less than a minute, John and Clayton had found another tick trail.

"I just can't believe them ticks have been on these animals," Clayton said, shaking his head and leaning back.

"And if they can do damage like you're talkin' about, John, well that ain't good."

"It's a concern all right," John said, and he handed me the last of the pelts that showed a blue trail under the black light.

I laid the last pelt alongside the other three on the table.

John leaned back onto his heels, stood up,
shut off his light and stretched.

"Well, I guess that's all of them—yup, four out
of the seven fishers show evidence of tick trails,"
he said and shook his head.

While Clayton is looking at the table, I look over
at Eric and nod.

"Oh! I forgot something.

Fellows, come on out to the cruiser for a
minute, will you?" Eric says turning and
opening the mudroom door to lead us all
outside.

We'd planned that little maneuver earlier too.

I follow Eric, hoping Clayton will follow me.

He does.

John hangs back, lingering over the skins as
any good scientist would.

But as soon as Clayton leaves the room, John
gets busy picking up all the marked skins and
sliding them into evidence bags he had hidden
in his jacket.

Eric walks over to the cruiser, opens the door
and then closes it.

With Clayton sorta sandwiched in between
the two of us—not crowded but surrounded—
I take a deep breath and break the news to
the beefy trapper.

"Clayton, there's a problem with some of your
hides.

We've been watching your traps all week.
Four of those animals that showed tick trails
have actually been marked with dye.

Those animals were found in traps that had not
been set in compliance with state regulation."

Clayton looks at me a little stunned.

I keep rolling rather than invite any comments
from him.

"These furs were obtained illegally and I'm going
to have to cite you and take those four fisher
hides as evidence."

John comes walking out of the cellar with
the black light in one hand and the skins all
bagged up.

He walks wide around us, pops the trunk and puts the evidence bags and black light inside and nods to us.

I'm writing Clayton a citation while Eric sorta stands back giving Clayton a "don't even think about trying anything" lawman look.

Based on Clayton's history, we are fully expecting Clayton to launch into some sort of rant and possibly take a swing at one or more of us.

But the big guy just stands there like he's thinking hard about something.

There's the usual scowl on his face, but it's a little softer. He looks puzzled.

It takes me a minute or two to write him up. Clayton just stands there watching me write.

He doesn't make a peep.

That's fine with me.

I hand him the paperwork and say, "You can pay the fine or have your day in court, Clayton. Up to you."

Clayton takes the citation from me like he is reaching for a full cup of tea on a dainty china saucer from his Auntie and nods.

Finally, he speaks.

"Okay," he says. "You got me. But what about those ticks—do I gotta spray my place now or what?"

"More than once Satan skidded our girls across the front lawn like water skiers as they tried to hold him back from taking off after a squirrel."

Satan
Blame the Name ?

Contributed by Richard Hislop

I grew up with dogs. And as much work or trouble as some people think they are, it is just second nature to me to have the sound of paws with claws clattering across a hardwood floor, a tail thumping on the living room carpet, a wet nose coming up under my palm from under the dinner table looking for a bite of my steak.

So, when our older dog met his maker, I went looking for another to keep me company out on patrol.

Because my territory covered a lot of Lake Champlain shoreline as well as smaller bodies of water, I figured a dog that didn't mind getting his feet wet would work best for me.

All my friends said labs make great companions and don't mind the water, so I went with their advice.

I picked this black lab puppy out of a litter for his big blocky head and stocky body. He was a purebred with a little white diamond on his chest—the rest of him jet black.

My goal from the get go was to have a well behaved dog I could take with me on patrol.

I don't know what possessed me to name this cute little Charlie Brown faced pup "Satan," but that's the name that came to mind when I went running through a long list in my head.

I guess I thought it was kinda tough guy cool.

I would pretty much be assured no one else was going to be yelling out "Come, Satan!" out in the field, right?

Lucky or Rufus or Beau or Max—there's thousands of dogs with those names out there, but Satan?

Looking back on it, I wonder if maybe there was a little voice in my head that kinda knew what I was getting myself into with this particular pup.

Other folks might say by giving a sweet little puppy a name like Satan, I brought my troubles on myself.

And then there's the folks who will argue all
the stuff that happened over the next decade
was nothing more than a coincidence.

I don't know.

I guess I'll just tell you some of my Satan stories
and let you decide.

I can tell you that right from get go, this pup
was a handful. He had a mind of his own.

At just a few months old, he figured out how to
get me to take him with me on patrol.

His trick was to make my wife, Carol, want to
kill him.

It started with digging.

Most labs don't much bother flowerbeds and
gardens.

Digging is a terrier's forte—something you
expect of small dogs bred to go after rats.

Once in awhile a Beagle will launch into
burrowing thinking there's a rabbit every few
feet beneath his feet, but labs just aren't
known for excavation.

But Satan loved nothing more than ripping into whatever greenery my wife, Carol, carefully planted about the yard.

So, from a tender age, every time he went outside to answer Nature's call, Carol had to worry about what Satan was tearing into and rooting up from spring to frozen ground for a dozen years.

I think he knew how much it annoyed her. And of course, if it annoys the wife, well, it is going to bug the hubby big time.

Listening to the two of us have a healthy discussion about his terrier tendencies over breakfast, lunch and dinner, Satan would just sit quietly looking all innocent.

But I swear he knew exactly what was going on and it was all part of his master plan.

Carol would remind me the reason I bought the dog was to have him go with me to work.

"He's your dog," she'd say to me. "He's your responsibility."

Tough to deny getting him was my idea.

Truth was, Satan grew out of cute puppy stage

and into a big brute real fast. He was more
than 85 pounds by his first birthday and topped
100 pounds by his second.

All muscle. Built like a line backer.

When he made up his mind to do something,
there was no way Carol or the kids could
control him.

He'd just put his head down and charge like
a stubborn spoiled pony. Stand in his way and
you'd get knocked flat. Hold onto his lead when
he was moving out and you were going to get
yanked off your feet.

More than once Satan skidded our girls across
the front lawn like water skiers as they tried to
hold him back from taking off after a squirrel.

Satan's lack of manners and grubbing about in
the gardens prompted the wife to make it clear
she was not interested in babysitting my dog.

She was already taking messages for me, watching
the kids, a couple cats, taking care of the house
and putting three good meals on the table.

I would come home to hear about how Satan
knocked something off a table or bowled the
kids over or his tail cleared a card table of

puzzle pieces the kids had been working on
for a week.

If I intended to keep this bull in a china shop,
I needed to take him with me.

That was the plan to begin with, right?

Well, I had to admit, it was. But Satan was not
like the calm companion I had before—a dog
that actually listened to me.

He was a blockhead. Outdoors, sometimes he
would come when I called him and sometimes
not.

It all depended on his mood. Satan had his own
agenda.

My first real indication Satan might never
understand his role as my partner was on a
clear, warm January thaw day in Hinesburg.

I decided to take him along for some ice fishing
license checks out on Lake Iroquois.

I figured it would be good exercise for him.
And even if he took off I would see him all black
against the snow—and he would come back to
me eventually.

What I hadn't counted on his was the dog's enthusiasm.

I parked the cruiser at the fishing access and opened the back door. Satan was ecstatic.

He jumped out and immediately set off smelling every bush and marking a goodly number of them.

I grabbed my binoculars, my citation book, checked my pocket for my pen, and reached for my ice chisel.

The chisel was rolled steel, a little more than five feet long, with a rounded handle at the top and a leather strap attached too, so I could slide my wrist through it. It had a sharp point at the end.

It was kind of like a ski pole or a hiking stick. Some folks might even call it a staff or a metal walking stick.

But it served a very specific purpose.

I used that chisel to test the ice out in front of me before I took a step.

I'd had a little incident the winter before out on Lake Champlain.

I'd gone through the ice more than a mile from
shore on a snowmobile with no one around—
about drowned.

I was following up on a complaint about dogs
running deer on Providence Island—up near the
Islands—when I hit a thin patch of ice covered
with snow near Carlton's Prize.

It was do or die time. No one to help you and
if you don't move faster than you ever have in
your life, you're dead.

I tell people that the fastest man in the world
could not have beaten me that day. I swam for
all I was worth to reach ice that supported my
weight.

But that kind of excitement does have an effect.
Some guys would've just turned in their badge.
I did consider that briefly.

But with the help of my supervisor, who told me
to soak in a hot bath, drink some whisky and
plan on snowmobiling a different piece of Lake
Champlain with him the next day, I got through it.

And now, when folks talk about "getting back
on the horse" I know what they mean.

But after taking that ice bath, I did search for

ways to do my job a little smarter. One of my tricks was to carry an ice chisel.

Swinging it out a few feet ahead of me, letting the metal nose tap the ice and listening, I learned to tell if the ice was thick enough to support me.

This day, I could hear Satan prancing and racing around on the ice and snow behind me —enjoying the day like the big puppy he was. He wasn't even a year old yet.

With the thaw underway, there was a glaze of water on top of the ice sheet and it was darn slippery footing. I was young, strong and maybe still a little naïve.

Experienced ice fishermen invest in crampons —ice grabbing metal clamps that affix to your boot sole with rubber or leather straps. These provide much better traction when traversing bumpy ice than any rubber soled boot.

But I didn't have any.

I saw a loose collection of ice anglers a couple hundred yards away and decided to start checking licenses there.

I swing my chisel out in front of me, listen for

the chink and clunks coming back to me and
step in a regular rhythm.

I'm maybe 30 feet away from the ice angler who
is busy jigging—just about to put my friendly
warden smile on my face and introduce myself
—when I saw the fellow's back stiffen, and a look
of total astonishment comes across his face.

They say you never hear the one that gets you.

I can tell you it's true.

But I did see something was about to get
me. The look on the fisherman's face told me
something was up.

But there wasn't a darn thing I could do about
it. It all happened so fast.

The guy's mouth falls open like a pouting perch
and his eyes open huge. I swear even the live
minnow on the end of his jig stops wriggling for
a second and stares at me too.

Whatever they are looking at is right behind me.

I start to turn my head and look but it's too late.

I feel a massive hit on the back of my calves.

My boots fly out from beneath me and I shoot up into the air and fall backwards like I'm being tossed in a blanket in some sort of college boy prank.

Only there ain't no blanket and there ain't no friends there to break my fall.

My ice chisel goes flying out of my hand.

I'm flat on my back in the air—looking straight up at the winter blue sky.

My arms are out at my sides.

I start flapping my hands in desperation—a 220 pound baby bird in a warden uniform.

Sadly, I'm a bird who hasn't fledged yet.

No feathers. No flight.

Gravity takes ahold of me and I fall back, slamming onto the ice—with my tailbone beating my skull to Mother Nature's blue cement by a fraction of a second.

Unh.

It all happens so fast.

The wind is knocked outta me. I'm flat as a stale gingerbread man at a New Year's party.

I just lay there flat on my back.

I suppose I look like a warden making snow angels for the entertainment of the ice fishermen.

I'm seeing stars. My brain is trying to figure out what just happened when I notice a thin dark missile heading straight at my head.

My chisel!

I wince and twist my neck to the left a couple inches—which is all the movement I can muster at that particular moment in time.

I hear a "Ka-thunk!" about three inches from my head followed by a musical saw "twaaaaaaang" that runs a good three octaves up the musical scale.

My chisel has planted itself upright in the ice like a spear just six inches from my ear.

I am lying there trying to get a breath, trying to coordinate my brain with my body again.

That's when I hear it: paws and claws on ice, some panting, followed by sniffing.

Satan.

There's no doggy apology, no face or hand lick.
If I'm dead, he intends to go through my pockets
for treats.

I hear the sound of boots with crampons
shuffling towards me and a man's voice.
Someone is talking. It takes me a few seconds
to make out the words but finally I get it.

"Oh my gosh! Warden, are you okay?"

I see a dark figure leaning over me and
extending a mittened hand.

I say, "Give me a minute, please," and work on
making sure the foot bone is connected to the
ankle bone and on up before I attempt to sit up.

I get a few good breaths back in me, then take
the fellow's hand and climb back onto my feet.

That's when I meet Joe—the shadow and voice
—and we have a proper introduction.

While I rub the back of my head, I ask him,
"What happened?" although I have a pretty good
idea.

"I looked up at you headed towards me and

behind you is this big black dog—barreling right down on you, running full bore," he says shaking his head.

"It was like watching a train wreck. Not a thing I could do to help you or stop him. There wasn't even time to warn you," he says, shaking his head in disbelief.

I thank Joe for his help, insist I'm fine and ask to look at his fishing license and his catch of the day.

His license was as fresh as his fish—he'd bought it just that morning before heading out.

"Thanks again, Joe," I say and walk over, pick up my ice chisel and turn to head on over to inspect some shanties. "I'll be on my way."

"Uh, Warden?" Joe says, stopping me in my tracks.

"Yes?"

"Just one more thing—about that dog of yours."

I'm looking at him in silence.

"You know he kinda had this big puppy grin

on his face as he was running up on you,"
Joe said. "He was probably just playing and
couldn't stop on the ice. But uh, just to be safe,
you might not want to turn your back on him
while you're out here."

I nod, smile and walk on—a little sore and a lot
wiser.

I learned a couple valuable lessons on Lake
Iroquois that day:

Ice doesn't care if you are 25 or 85, and

Never turn your back on Satan.

*"I stood there kinda
dumbfounded by the situation.
My dog was out fishing.
He had already moved on,
made new friends."*

Satan
That'll Teach Him

Contributed by Richard Hislop

One of Satan's worst traits was his refusal to wait until I got the boat tied up before getting out.

Slimy water, belly deep mud, reeds as thick as teeth, it didn't matter.

He would start an antsy pants dance and tail wagging and whining and drooling as soon as I pointed the boat towards shore.

Throttle back, get within 50 feet of land and he would leap like a deer clearing a 10 foot apple orchard fence, slam into the water, bob up like a cork and start paddling madly for shore.

If Satan were rescuing someone from drowning, his act would be really impressive. But he weren't no Newfie.

Satan just wanted to check out the scenery.

In fact, if someone had been in the water and
needed assistance, their only hope would be to
grab his tail as he went swimming by.

In all probability he would slam a swimmer into
unconsciousness if one were so unfortunate
as to be in his way. This dog was never big on
considering others. Built like a bull with a mind
to match, you know what I mean?

All that advice I got about getting a water dog?
In retrospect, I probably would have been better
off with a dog that didn't like the water so much.

If I planned on being anywhere near water with
Satan, I had to plan on being wet. He could
make water part like a Sea World star.

It didn't matter the season either—Lake
Champlain at 45 degrees in March or a balmy
90 degree day in July—it was all the same
to him.

The fact he soaked any human standing nearby
—that would be me—meant nothing to him.

He didn't have to keep his uniform impeccably
clean to greet the public.

All Satan had to do was shake and sit in the
sun a few minutes to dry off.

I thought of tying him inside the boat, but I was
pretty convinced he would just jump anyway
and if he was tied, he might hurt himself.

Satan also had a tendency to paw and claw—
remember his digging skills in the garden?—
and I didn't want my boat torn up either.

I tried a big selection of voice commands. Some
of which can not be printed here. It didn't
matter. Satan pretended to be deaf.

And to be honest, it wasn't just the leaping
out of the boat that bothered me—although
having people look at you like your boat must be
on fire would be embarrassing enough for most
people.

Once on the beach, Satan would frequently
disappear. He could be gone for anywhere from
a half hour to a full day.

Just checking things out, I guess.

I'd hike, track and holler.

Nada.

A heck of a way to impress members of the
general public who might be wanting a little
piece and quiet, right?

I got sick of his behavior and decided it needed to change.

I consulted with dog trainers and anyone else who would listen and was advised the best thing for me to do was just leave him—literally get in the boat and leave the area.

Once the dog realized he was alone—Daddy was gone—he would have some sort of canine epiphany and never leave me again for fear of being left behind.

That was the theory anyway.

In retrospect, maybe the guys who told me this had some abandonment issues of their own.

I mean, why would someone believe dogs consider the consequences of their actions? Heck, a lot of humans don't even do that.

But then, we often expect more from our dogs, I guess.

For want of a better idea, I decided I'd give their advice a try the next time Satan launched out of the boat.

I didn't have to wait long.

It was a lovely July day on Mallets Bay in
Colchester—the kind of day when you get to
enjoy some of the best weather Vermont has
to offer.

I had Satan along for company mostly because
we had just planted a big garden. If Satan dug
into it when no one was watching, he and I
would both be in big trouble.

The dog did pretty well the first couple of hours.
I checked a number of anglers trolling the
broad lake, then I headed into the mouth of the
Lamoille River to get out of the noonday sun and
stretch my legs.

I figured there would be anglers in smaller craft
wetting their lines here as well as some shore
anglers whose licenses I could check.

As I rounded the bend, I made a mental note
of the number of people in each boat and how
many were fishing. Among the half dozen
boaters I saw was a pair of older guys in a
painted aluminum rowboat.

I was maybe 200 feet from land, aiming for an
open dock when Satan started whining. I knew
he was winding himself up and getting ready to
pull his same old trick.

I told him to sit. He ignored me.

It was clear to me he was in another one of his "Sigh-uh-naw-ruh, Sucker" explorer modes and not about to listen to me.

Satan began scrambling his paws on the deck, ran to the bow, looked off the port side, then dashed to starboard, then back again.

When I slowed to a crawl and was within 80 feet of shore, he jumped up on the rail, perched like a cat on all fours for a second, then leaped big and started swimming towards shore.

I cut the engine and watched him. Head high, tail out like a rudder. He swam like a dog possessed. I had to give him credit for great swimming skill.

When he got to shore, his nose shot up and down like one of those wacky duck lawn ornaments and his tail waved back and forth like a dowser's divining rod. He didn't even look back at me to wave goodbye.

"Okay, Satan. Time for you to learn a lesson," I muttered. I spun the wheel hard, pointed the bow back to the open water and hit the throttle.

Heeding the advice of the experts—I left my dog on the shore.

I guided the boat to another piece of the bay and continued to check licenses and inspect boats while keeping one eye on my watch.

I figured an hour—maybe two hours tops—should do it.

I had visions of Satan sitting on the dock looking out at the bay, all contrite and apologetic. Or maybe I'd find him on the beach pacing and pining for me.

Then again, he was smart. Maybe Satan would scratch an SOS in the sand with his claws.

I just really hoped I would find a new dog waiting for me when I went back to get him—a dog that listened.

I forgot who I was dealing with, I guess.

I rounded the bend and looked out to where my big black blockhead buddy jumped ship less than two hours earlier and saw—nothing.

Hunh.

Well, it could be he was just down the shore a

little bit hunting for me. So I slowed the boat
some more, came in closer and looked harder
and farther up and down the beach.

Still no sign of my dog. Now, I was getting a
little nervous.

The sun would set in another couple hours.
I had been on the water all day, my sandwich
was long gone, my neck was sunburned and I
was ready to head home. I didn't want to spend
half the night hunting for my dog.

I pulled up to the dock, threw out my fore and aft
fenders, shut the motor off, tossed my lines around
the cleats and walked on over to look for dog tracks.

That's when I heard a yelp coming off from
behind me, on the water. Only the yelp wasn't a
dog's it was a human's.

I spun around and looked out over the bay.
The sun was shining bright on the lake and it
was a little tough to see. There were a couple
Starcraft runaboats and a no name older trihull
and then off over in quieter water, the same worn
aluminum rowboat I had spotted hours earlier.

The two old guys holding fishing poles hadn't
moved far from their original location. Must be
the perch or croppie fishing was good there or

they were too lazy to pull up the anchor.

Was one of them doing the yelling?

Were they in trouble? Something about them was different.

I looked harder and saw three people in the rowboat now. Someone shorter—maybe a kid— was with them. No fishing pole out, just sitting.

But something wasn't quite right. The shape of the third person and the way they were sitting looked kinda off. No shoulders, their arms at their sides and kind of a strange head.

I lifted my field glasses to my eyes. There was a blocky black head and long ears smiling up into one of the fisherman's faces.

I stood there kinda dumbfounded by the situation.

My dog was out fishing. He had already moved on, made new friends.

He wasn't sitting there on the beach a nervous, panting, drooling wreck crying and begging for his master to come back.

In fact, as I watched, I saw one of the guys was

feeding Satan pieces of a big fat sandwich.
The other guy was laughing and handing my
dog what looked to be potato chips.

Oh yeah, Satan is suffering all right.

His neck wasn't sunburned and sore. And obviously
his stomach wasn't growling like mine was.

Now, the fellow in front sees me watching them.
I'm holding up my binoculars. He waves, yelps
and his buddy pulls up the anchor.

Within 10 minutes they are maybe 100 yards
from pulling into the dock opposite my boat.

I'm envisioning having to jump in the water to
steady their craft as they approach the shore.

I'm certain Satan will be head over heels with
puppy dog love once he catches sight of me—his
beloved master—waiting for him on the dock.

So, I'm big time concerned the dog will do what
he always does—rocket off the bow. And if he
does that to these older fellows in their little
boat, he'll likely dump them and all their gear
into Malletts Bay.

I stand on the end of the dock ready to leap in

the water up to my waist or even swim out to
these guys if I have to.

Who knows if they can swim? If they even have
life vests in the rowboat?

I watch Satan.

He sits quietly as a statue in the front of
their boat as they motor in. No ansty pantsy
squirming, wiggling or yipping in front of his
new best friends.

He's a different dog. He remains seated all the
way into the dock. Regal as a lion.

Only when the first old gent has creaked and
moaned and gotten himself out onto shore and
the second fellow tells Satan, "You can get out
now, Boy," does my dog gingerly step out one
paw at a time.

They could have left an open thermos full
of coffee on the front bench seat and Satan
wouldn't have spilled a drop.

Amazing.

He looks up at me and smiles in a quietly
triumphant way.

He's playing this for all it's worth.

I start to open my mouth to introduce myself.
But they cut me off.

"Say, if you don't mind me asking there,
Warden, just what in the world were you doing
dumping this fine dog on the beach and then
going off without him?" the first old fellow
asks me.

He had a look in his eye that said if he were 40
years younger he would just punch me in the jaw.

His eagle glare gave his fishing partner courage
and as soon as he was able to stand upright on
shore, he decides to take a bite out of me too.

"Yeah, what are you—a dog hater or sumtin?"
he says and spits into the lake behind him.
"You oughtta be ashamed of yerself for treatin'
him that way."

"I was only trying to teach him a lesson," I
replied, "You see...."

"A LESSON?" they squawked in unison.

With that one little word, all hell broke loose.

I felt like a French fry pack spilled at the

McDonald's on Shelburne Road in front of a dozen hungry seagulls.

There was a lot of noise and nowhere to hide.

The pair started flapping their arms and dancing on the shore just like the birds battle for fries —screaming louder and louder.

I couldn't get a word in edgewise.

"Teach him a lesson?!" they both screeched at once.

"You should be reported to somebody or other. This is one fine dog...." the first fellow said giving Satan a pat on the head.

"Yeah, what is wrong with you?" the second fellow said, spitting again into the lake.

The two of them were working themselves into a frenzy. They waved their arms, muttered and screeched all around me.

Maybe it was part of their athletic program—a kind of senior calisthenics—a way to get their blood flowing to prevent a blood clot after sitting in that little boat all day or something.

Every time I opened my mouth to defend myself it would just wind these guys up more.

They were like a wrestling tag team—as soon as one ran out of breath, the other one jumped in with more ways to question my character.

I give up on trying to explain my actions and look down at Satan.

I detect a twinkle of delight in Satan's eyes as he sits listening to his fishing buddies rant. Maybe he knows they are chewing me out.

Satan's mouth falls open in a big doggie smile and he runs his sloppy red tongue around his open maw. I can't tell whether he's tired or laughing at me.

It's time to go home.

"Thank you gentlemen for taking such good care of my dog," I say stepping across the deck to loosen the lines.

I jump down into my boat, pull in the fenders and nod to Satan. He stands up, trots onto the dock and leaps into the boat beside me.

I'm not sure if he really wants to come with me or it's just that he never passes up the opportunity to go for a boat ride. I think maybe the dog figures he's made his point and I've suffered enough.

Even with Satan in my boat, the old gents still continue to squabble.

"You sure you don't want to just leave that dog here for good? We'll take him," one of the men says. Now, it's my turn to pretend I'm deaf.

I turn the key in the ignition, untie the bow and stern lines from the cleats, throw the boat into reverse and quickly back my way off the dock.

The fellows on shore are still flapping their gums, kicking up dirt and pointing at me.

It is sweet relief to spin the wheel, plunge the throttle forward and hear the boat motor roar to life. The noise drowns out their squawking.

I look down at Satan. He smiles at me, then runs to the bow and turns his face into the breeze.

"That'll teach him, they said," I shout into the wind and laugh.

"Yeah, right. Teach who?"

"When the angler finally unleashed his cast and let the big spool of line sail forward, Satan exploded."

Satan
Dog Fish

CONTRIBUTED BY RICHARD HISLOP

I had Satan with me checking licenses out on the Winooski River not far from the city that gives the river it's name, above the old mills.

It was the middle of May—a perfect late spring day and I knew a Hendrickson hatch was on. Serious fly fishermen—and women—would be out on the river putting all those flies they'd tied in January to use.

Satan was like a colt in the spring about ready to bust down a stall door to get out onto some green pasture. He began pawing at the passenger door as soon as I slowed to pull off onto the gravel parking area.

Whether it was Spring fever or just youthful exuberance, something in the air was calling him.

Once out of the cruiser, he raced down the path
and plunged into the cold water and began
lapping up frothy mouthfuls like it was ice
cream.

I couldn't help but smile to see him so happy.

He was living in the moment and enjoying life—
having a grand old time racing into the stream,
getting his paws soaking wet and grinning back
at me, waiting for me to throw him a stick.

Thing was, this was a grass and hollow reed
only bank. There was no stick around to throw
him—not even a sumac bush nearby to bust a
branch free.

Being his impatient self, Satan looked for
alternative amusement.

He swung his head left and then right and then
spotted a fellow about 100 yards downstream.

I followed Satan's stare. The fellow was outfitted
in fine chest waders, a tan vest with bulging
pockets and was wearing a crushable hat with
a good brim to help keep the water's glare out
of his eyes. He looked to be serious about his
gear. I figured on going to introduce myself and
checking his license in a few minutes, once I got
Satan calmed down.

He was moving his right arm back in a perfect fulcrum arc—slowly and methodically like a metronome—the classic precast swings.

As a budding fly fisherman myself, I stopped for a second, to admire his technique.

This guy was good.

Of course, what I saw was not what my dog saw.

What Satan saw was a new friend just down river teasing him with a long skinny stick— getting ready to throw it for him.

In true Satan style, he took off like a rocket towards the object of his desire—a human who wants to play with him.

I call, "Satan! Come here, Boy! Satan! Come!!," with my voice getting louder and louder as the dog runs farther and farther away from me.

I might as well have been shouting for the river to stop flowing.

Satan had turned off his ears.

He was gone.

"Darn dog," I muttered to myself and shook my head in disappointment.

I started a slow trot after Satan. I knew there was no way I was going to outrun him and I could easily bust an ankle on loose cobbles hidden beneath the bubbling foam and moss if I ran full tilt in a futile attempt to catch him.

I was formulating an apology to the fly fisherman in my head.

Nothing an angler hates more than having some dope spook the fish he is trying to entice with an au natural presentation of his flies. And by the looks of this fellow, he was serious about fly fishing, probably tied his own flies.

And here was my mutt—the warden's dog no less—racing right up the middle of the brook, scattering every fish in the area.

This fellow wouldn't be able to find a fish on this section, once Satan got done plowing through it, for a good half hour or more.

The guy—and maybe the fish too—should write Satan and me a ticket for disturbing them.

Oh well.

I was a good 50 feet from the guy when I saw it all happen in slow motion.

The fly fisherman was turned upstream. With the noise of the water spilling over the rocks and his total focus on fishing, I realized he had not yet seen Satan or me. He was only watching for signs of fish and where his fly was going.

This guy had no idea there was a 110 pound black lab barreling down on top of him like a bear cub.

I stopped dead in my tracks and held my breath waiting to see what Satan would do.

I didn't know if my dog was going to knock the fellow over or sit down and shake hands or just what all he had in his demented doggie mind.

I only knew I couldn't do a darn thing about it.

It was the kind of suspended in time feeling you get when you are in your car watching a fender bender unfold in front of you in the parking lot of a supermarket or watching a vase get bumped by a guest's elbow at a party, crash to the floor and shatter into a gazillion pieces.

Even if I honked my horn at the drivers or dove across the living room to make a flying catch for the vase it wasn't going to change the outcome.

I was helpless to stop Satan.

But I never in my wildest dreams thought Satan would do what he did.

Stunned, I watched as the fellow reached for a big pool maybe 30 feet away and slightly downstream. He was on the third and final arc of a big back cast with the line coiled artfully against the blue sky behind him.

I saw Satan's eyes go wide following that line. The dog stopped short for a split second to watch—his mouth was wide open and his head rocked from side to side like a tennis player waiting for the big serve from his opponent at Wimbledon.

It was all a great game to him. He was waiting for the guy in waders with that brown long skinny stick to throw it.

So far, his new buddy just kept teasing him with half halts and back casts.

When the angler finally unleashed his cast

and let the big spool of line sail forward, Satan exploded.

I guess the tension was too much.

Since the guy hadn't let loose of the stick in his hand, Satan went after the next best thing—the skinny string with the fuzzy thing at the end.

He blasted through the water.

Only then did the fly fisherman finally see Satan. And by then it was too late.

Satan galloped over the rocks and water like a grizzly cub chasing a stray salmon through the shallows of the Kenai River in Alaska.

He was running alongside the line waiting for that little nappy thing on the end of the leader to fall close enough for him to make a grab. He was prancing in delight.

When the fly was about four feet above the pool, Satan leaped into the air and wriggled his body upwards to get more air, grabbed the guy's fly and bit down big.

He twisted his head like a bass grabbing a wooly bugger—mouth open wide—straightened his body as gravity pulled him back to the river and landed with an artful four footed splash that would have made an osprey smile.

If I had not seen it with my own eyes, I would never have believed it.

My feet stopped dead in the water for a second time and my mouth fell open to my shirt collar.

My dog had just grabbed this guy's fly.

How crazy is that?

Did the dog think it was the world's smallest Frisbee?

Who knew Satan could even see that well?

Who knew Satan was that coordinated?

Who knew Satan was that stupid?

I shook my head unable to believe my eyes, leaned forward and squinted hard trying to block out some of the glare from the sun on the water.

I looked back over at the guy with the pole —thinking maybe I am making all this up and what I just saw did not happen.

I saw the Orvis clad angler's jaw fall open like he was a trout about to reach up and swallow a big bug himself.

That was confirmation I was not hallucinating. I hadn't lost my mind.

Satan had indeed done something truly outrageous.

I saw the guy slam his rod tip to the water and peel off line like it was a lit fuse and he was holding a live bomb at the end of his line.

He was doing his best to do my misguided mutt a favor and take all the pressure off the hook in Satan's mouth.

That was nice, I thought, considering the fellow had just caught what was in all probability a Green Mountain first—a big black blockhead weighing in at 110 pounds—Vermont's first dog fish.

I picked up the pace and ran towards Satan.

He was standing in the stream with a decidedly different expression on his face.

He looked like a kid who's been running around the house chasing his bigger brother, slips on the hardwood floor, slides long, takes out the coffee table with his head and splits his lip.

All that was missing was mom yelling from the kitchen, "I told you boys someone was going to get hurt!"

My dog was shaking his head slowly back and forth—like a bumble bee had stumbled into his huge mouth and he couldn't shake it loose.

Forget trying to keep my feet and pant legs from getting soaked. I jumped into the pool beside Satan to steady him and knelt down.

My right hand reached for my hunting knife and my left hand went for his collar.

I had no idea where Satan had hooked himself taking the guy's fly, but I was figuring on having to do a little stream bank surgery.

I was grateful the dog hadn't taken off with the hook in his mouth.

The way he leaped and took the fly, maybe some swordfish had slipped into his gene pool from way back when, but it looked like there wasn't any salmon—he didn't run.

Satan let me get a hold on him.

He had kind of a confused look on this face—a mix of a scowl and guilt and a little pinch of pain. He looked up at me and his eyes showed some white.

"Hold still, Boy. Let's take a look," I said to him quietly.

I reached down to his floppy jowls, lifted up and began my inspection.

I heard the squish and rub of rubber boots splashing up behind me. I turned and saw it was the fly fisherman.

"I'm sorry," the fellow said, "I didn't see him until I'd cast and he was right on my fly."

He'd left his pole on the shore and was running the line through one hand and carrying a Leatherman tool in the other as he approached.

Seeing I had Satan settled down, he dropped his line and used that hand to pull a pair of librarian type fold up magnifying glasses out of his left breast pocket.

He slid the spectacles onto the bridge of his nose and leaned in and said, "May I help?"

I leaned back on my heels a bit and took a

harder look at the guy: salt and pepper hair, good hair cut, moustache, gray blue eyes, wearing an expensive vest and waders, lambs wool holding a nice selection of flies on his vest and all of it topped by a floppy khaki hat to keep the sun out of his eyes. A late fifties or early sixties model of a man.

He'd invested some serious money into his hobby. Probably been at it a good number of years.

I looked at his hands. No stains, no calluses, no missing digits. Clean unbroken nails. He had the hands of a guy who works behind a desk or maybe a surgeon.

"A surgeon would be real good right now," I thought to myself as he lifted up Satan's lip with the closed edge of the Leatherman.

Guess he had a liking for his fingers too. Maybe that's why he still had 'em all. His move showed a high degree of common sense in dealing with a strange dog. Some folks would just plunge right into the tiger's mouth.

Since he had the tools already in hand, I decided to let him play doctor.

Heck, those fishing vests carry so much stuff I

wouldn't have been surprised if he'd pulled out a folding table to do this oral surgery and an 10,000 watt operating room lamp.

I quietly slid my Bowie knife back into its sheath and felt a bit inadequate. That blade was designed for whacking saplings not fine work like this.

"Hold his head up just a little higher, please, and to the left?" the stranger said.

Together, we inspected one side of Satan's black droopy jowls and then the other and didn't immediately see anything obvious.

Hunh.

But this was a classic case of follow the leader —in fly fishing that's the name of the narrowing monofilament line that leads from the darker line to the actual fly that fools the fish—and the occasional dog.

"Ah! I think I see it," he said, lifting Satan's head a bit higher.

Satan quivered and I gave him a pat. He was being surprisingly cooperative standing here in the middle of the river with ice water running under his belly.

"Open up, Satan," I said.

I bent down to see what the fisherman saw and there it was—a hook with about three to four inches of thin monofilament leader hanging from it.

It was lodged a few millimeters behind an incisor, stuck in his gum line.

"What kind of fly was it?" I asked.

"That's a number 10 with a Muddler Minnow," the fellow replied. "Looks like it missed his tongue entirely."

I was new to fly fishing, but his fly rang a bell. My recollection was it could be fished on the surface or beneath and was comprised of turkey quill and deer hair.

I chuckled to myself, "Deer and turkey—two of Satan's favorites. No wonder he made a grab for this thing!" Of course, there was no way the dog could have known what it was made of. But these were critters he had tried to chase.

"Oh yes, I see it—not too far back," I said to the doggie dentist. "Can you get it?"

"I think so. I file down the barbs on all my flies," the fellow said. "It should come out easy. Just tilt his head a little more this way and keep his mouth open wide."

He adjusted his grip on the Leatherman tool and went in for the grab.

"Say, Aaaaah, Satan," I said trying to lighten the mood a bit and distract him.

Satan's eyes went big and white as I spread his jaws a bit wider and lifted his tongue and pushed it aside.

The doggie doctor slid his hand above and in between my fists, put his pliers on the shank of the fly, gave it just a tiny bit of side to side action to loosen it, then bent the nose of the pliers up and lifted.

Out the muddler came.

"Got it!" the fellow said with a self-congratulatory grin and held it up about 18 inches in front of my nose.

I let Satan's jaws go shut and gave him a few pats on the neck. Then I stood up and clambered out of the pool—it was a little icy in there.

I took a closer look at the fly—what was left of it.

It was basically a bare hook with a saliva soaked splash of gold wrapping around the shank. There was a tiny bit of hair and what once might have been feathers, tied to look like a wing.

I looked down at Satan. He was running his tongue around in his mouth, like someone just leaving the dentist's office after having a cavity filled.

He shook himself hard and wagged his tail.

I knew Satan would be fine when he dipped his head down to the river and took a few big gulps, then trotted off following his nose upstream.

I introduced myself at last and added, "You seem to have a pretty good technique with those pliers. Can I ask what do you do for a living?"

"I'm a dentist," the fellow laughed. "But I have to admit I've never tackled a dog before. Name's Jim," and he put out his hand.

I shook the fellow's hand and marveled at Satan's good luck.

Then I started to apologize in earnest.

"Uh, Jim, I'm really sorry my dog ruined your fishing here. I—" when he stopped me short.

"No need to apologize, Warden," he said. "Heck, I can fish and catch trout on most any decent day. But a big black lab taking my line and running with it?" he paused, shook his head and grinned wide in amazement.

"Well, that sure doesn't happen to many of us, does it?" he laughed.

I had to agree with him there. I nodded and smiled.

I hadn't really expected him to be so good about Satan ruining his fishing.

Jim looked over at Satan, who was off sniffing near shore now.

Then he went on. "Seriously, I'm glad your dog is all right. From the looks of him, he's already forgotten all about it."

Jim pulled a white handkerchief from his front pants pocket, wiped each of the Leatherman's pincers clean, then folded the tool up and slid it back into his vest. He did the same with his handkerchief—folding it neatly once again and sliding it back into the same pants pocket.

"Nope, I think I should be thanking him and you."

Seeing the confusion on my face, he went on.

"You see, my fly fishing club down in the Berkshires has an annual banquet in January and, of course, we all swap fishing stories of our previous year's adventures."

His eyes twinkled brighter. He looked like a guy holding the only winning ticket on a dark horse nobody else had taken seriously.

"The best true fishing tale of the year gets the member's name engraved on a trophy that stays within the club," he explained.

"I've always wanted to win it, but to tell you the truth, my life isn't very exciting. Other than getting some nice fish on the line now and again, well, nothing much has ever happened to me out here."

"But today, you and your dog just gave me a story I can't imagine anybody's going to be able to beat!"

Jim paused and looked off. He had a look in his eyes that said he was already there—basking in the glory of a well told tale—his pals laughing big, shaking their heads in disbelief, slapping him on the back and buying him another snifter of brandy.

"Yes, sir. I can't wait to share this one!" Jim said proudly.

"Massachusetts, hunh?" I thought to myself.

Well, that's a little bit of a break. At least he's not local.

If I'm really lucky, maybe by January, Jim will have forgotten my name too.

"If we had a fightin' chance when we caught the boulders, once the dog stood up and started his spin, and I couldn't see to attempt to coordinate my strokes with Herb, we were doomed."

SATAN
STAY

CONTRIBUTED BY RICHARD HISLOP

S atan mighta been a little off in terms of how his brain worked, but he was no dimwit in the dexterity department.

Take the time I teamed up with neighboring warden Herb Conly for a half day of license checks canoeing what the locals call the Huntington River over in Jonesville off of Route 2.

The Huntington is really just a stretch of the much bigger Winooski River Basin leading to Lake Champlain and on out to the St. Lawrence Seaway and Atlantic Ocean. But call the river what you will, it's popular with bait and fly fishermen from Spring through late Fall and has some lovely long lazy stretches for canoeists to traverse.

Spring run off, heavy summer downpours and ever shifting gravel bars can make the river a little challenging, but in late spring and early summer, once the water settles, it's a piece of cake.

Or so I thought.

I'd invited Herb along on this Saturday in June knowing there would be a good number of fishermen who had heard the river had just been stocked with fish. They'd be eager to try their luck.

Herb and I would check licenses and creels in style. We could cover a lot of territory this way as opposed to bulling our way through the brush along the road looking for anglers one at a time.

It's also a lot easier to get the job done with two folks in the canoe. The warden in the stern steadies the canoe while the man in the bow asks for the angler's license and checks their creel.

I don't know what possessed me to think Satan would be an asset on this particular patrol.

Maybe it was the glare Carol gave me as I was about to head out the door without my dog that convinced me to bring him along that particular morning.

She saw me bundling my gear together and Satan sitting at the front door wagging his tail.

He was looking up at me with that, "Let's go!" look but his lead hung limp on the coat rack.

Carol shoulda been a detective. She caught me right in the act of leaving Satan behind.

"You're taking your dog with you today, aren't you, Dick?" she asked innocently.

"Uh, I'll be on river patrol with Herb today, Dear. I hadn't planned on taking Satan along," I replied.

"I sure wish you would," she said. "I've got to be out running errands all day. I don't think it's wise to leave him here."

Well, the wife had me there.

Who knew what damage Satan would do if left to his own devices home alone.

Put him out in the kennel and he might just dig his way out and take off, or worse, stick around and tear through the flowerbeds like a bulldozer.

Leave him in the house and you might lose a chair leg to his chewing.

This was one of those "devil or the deep blue sea" choices.

I looked down at Satan. He was sitting quietly and looked up at me as innocent as a normal dog. He wagged his tail—just once—again. Politely pleading.

He sure was well behaved this morning.

Always the optimist, I gave in.

"Okay, Satan, let's go," I said, and I picked up his lead and snapped the clip to his collar.

I met Herb 45 minutes later over at Bolton Flats.

Of course, when Herb saw Satan sit up in the truck, his eyebrows shot up and his welcoming smile turned into a tight lipped line.

My dog had been preceded by his reputation. I was hoping Herb would give Satan the benefit of the doubt this day.

"Your dog coming with us?" he asked with some concern his in voice.

"Yes," I said. "He'll be fine. He generally lies quiet in the bottom of the canoe."

That was a truthful statement. Satan did generally lie quiet—until he started scrambling and whining and winding himself up like a spoiled three year old kid in the candy aisle.

But I had a plan to nip any temptation Satan might have right in the bud.

I figured I'd sit in the stern with Satan right in front of me on the bottom of the canoe.

Even if Satan did start to wiggle and sit up, I could grab his collar and steady him or pull him back down onto the bottom of the canoe with a hand if my voice command didn't work.

The water was low. The river was lazy. This should be a sweet, quiet ride.

It was a good theory anyway.

What I hadn't given a lot of thought to was all the weight we were asking this bunged up tin can canoe to carry.

I'm a generous 225 pounds and Herb is a good 170 pounds and then Satan, he's another 100 plus pounds and then there's all our gear—a big cooler filled with a couple big bags of ice, our water and lunches, radios, life vests, paddles, etc, etc.—all crammed into a 15 foot aluminum canoe that had been dented, dinged, welded, pounded on and pounded out for probably 25 years by countless wardens before us.

Heck, maybe even Hiawatha sailed this one, I don't know. It was one step above rotten birch bark.

We scrambled down the bank and loaded all our gear in, then I climbed in and walked the keel line to the stern.

Satan was next.

I said, "Come on, Satan. Let's go," and he stepped in as lightly as the handsome prince not wanting to wake up his princess, sat down facing me and gave my chin one long lick.

The thing was, I had forgotten how big my mutt was.

It looked like I had invited a black bear cub into the canoe,

But all good, we were just gonna make this work.

I breathed a quiet sigh of relief that Satan
was behaving so well and looked up at Herb
and beamed like this is the way my dog acts all
the time.

Herb smiled just a bit and nodded in approval
despite himself and grabbed the line, readying to
push us off into the river.

"Okay, Satan, lie down," I said, "Lie down."

He looked at me, licked his chops,
and did it.

Wow.

This trip was off to a good start.

"You ready?" Herb asked.

"Yup, come aboard," I said with hearty
enthusiasm.

This was gonna be some fun.

But when Herb got in, what was six inches of
clearance along the gunwales dropped about in
half.

We were barely above the waterline.

I scowled and thought to myself, "I probably shouldn't have had that second piece of coconut cream pie last night."

As if that woulda made a difference.

I was hoping Herb wouldn't notice how deep in the water we were buried.

Being lighter and in the bow, he sat higher.

I plunged my paddle into the water and pushed back, pulling the canoe loose from the bank.

Herb dipped his paddle into the current and we turned downstream for a good morning's work.

We were off.

But Herb immediately spotted the problem.

"We're kinda low in the water, aren't we, Dick?" Herb asked about 20 feet from shore.

"Yeah, a little," I replied. "But we should be all right as long as we keep her straight," I said —trying to convince myself maybe?

We rounded a few bends in the river and floated up on four different anglers trying their luck.

Between Bolton Flats and Richmond, Herb and I checked a good dozen licenses and issued a couple warnings.

We'd been out on the water a good three hours and were doing real well until we approached the narrows near the Jonesville Bridge near Duxbury.

There's a reason the engineers chose that spot to put the bridge—less concrete and steel.

But with that kind of bottleneck comes gravel bars and swifter currents as the water and debris all tries to run through the funnel at once.

We'd been doing so well.

I was looking off at the shore inspecting an otter slide when Herb said, "Dick, you see what's ahead?"

I dipped my paddle a little deeper and twisted the face into the water, to turn the stern out for a better view.

One peek was all it took to know we knew we were in trouble.

Boulders, driftwood and sinuous gravel bars

were all mixed up with deep swirling pools of dark green water.

It was a minefield for an old bottom dragging tub like ours. There was no clear path down the middle.

I felt the canoe being pulled forward.

"We might scrape bottom a bit," I said to Herb, "Be ready."

"Okay," Herb said and he reached out farther with his paddle to help straighten the canoe.

That's when the tub scratched over a couple of boulders set in between two deep pools.

Mother Nature had cleverly hidden 'em beneath about 10 inches of white water.

Trouble was, we were sitting a good 12 inches deep in the river.

The aluminum bottom of our canoe shuddered, shook, heaved and screeched like a serrated knife on a metal roof, scraping across those rocks.

Herb—being lighter and with the benefit of my

bulk in the back lifting him in the bow out of the water a bit—made it over. But when the canoe approached Satan and me, it sounded like two desperate teenagers were let loose in a metal shop.

There couldn't have been more noise if they'd turned on every machine in the shop. No earplugs either.

The river was ripping into our canoe bottom like a kid pounding on a loose roll of metal flashing —bashing it with a ball peen hammer while the other tried metal sheers.

Satan—who had been sleeping soundly— rocketed about two feet into the air.

The hair on his shoulders was straight up. His mouth was open wide in shock—and his eyes were about coming out of their sockets.

He spun in the air to face me—looking for an explanation from the top dog—me.

Satan's tail smacked Herb's head like a 2 x 4, laying him over towards starboard and temporarily disabling him.

The canoe bottom is rippling and buckling

beneath our feet. Satan dances as if he is on hot coals in reaction.

I can't see anything but black fur, white teeth and red doggy gums.

I dig my paddle into the gravel bank hoping it will be enough to push us up and over the gravel bar and back into water.

My canoe paddle bows out as it catches on a combination of gravel and river rock. I push harder. I've got nothing to lose.

If I had to pole vault this tub or break a paddle to get us back in the water, so be it.

The canoe groans and shudders and slows even more. Satan spins in a circle like a roulette wheel. Now, I get slapped silly with his bristletail.

Suddenly, he stops his spin and gets right up into my face looking for answers.

A giant red tongue and pearly white fangs are planted in my face. I suck in a big gross dose of hot dog breath—my only source of oxygen.

I wish for an extra pair of hands, to push Satan

down and out of my way so I can see what Herb is doing in the bow.

I feel like the guy who sticks his head into the lion's mouth. Worse.

At least that guy gets to turn his head sideways and breathe fresh air.

I am staring straight down Satan's gullet and sucking in kibble fumes from three days ago.

Screech. Grind. Groan.

It's like we are sitting in a tin can being dragged down an asphalt highway.

Unable to see Herb and balance the boat, the canoe starts to tip.

"Stay!" I shout to Satan.

But it's too late.

I feel water spilling onto my lap. It's coming over the gunwales.

If we had a fightin' chance when we caught the boulders, once the dog stood up and started his spin, and I couldn't see to attempt to coordinate my strokes with Herb, we were doomed.

I can't give you a blow by blow because Satan had swallowed my head, but I know Herb was doing everything he could to keep us upright.

All I can say is, we were goners.

The canoe flipped over like an outta balance hammock—dumping Herb and I and all our gear into the water.

It might have been the middle of June, but the water felt like March. It was just a couple degrees short of ice cube making in the deep pool I tumbled into.

There were some pretty strong currents roiling up inside that little cauldron too, which might also have contributed to our inability to keep the canoe upright.

When I came up to the surface, I looked around for Herb, Satan and our canoe.

I found Herb about 10 feet away from me—he was clawing his way up on top of the gravel bar, looking back at me and scowling like a cat who

just suffered the indignity of a slip into a full
bathtub.

I swam a couple short strokes, got my feet
beneath me on the rocks inside the pool, stood
up and climbed on top of the gravel bar to stand
beside Herb.

I was shaking the river out of my hair and eyes
and looking for our gear and canoe.

I saw some of our stuff floating around—hats,
duffle, a small cooler—but I didn't see the canoe.

Where the heck did it go?

Did it sink?

That didn't make sense.

I should see a bit of it sticking up somewhere
along shore or stuck on another pile of rocks.

"Where's the canoe?" I asked Herb, who was
shaking a pant leg—trying to wring some of the
water out of his trousers, while checking his
pockets to see what he'd lost to the river.

Herb gave me a look that was half disgust and
half pity. He sighed, stood up straight and

pointed downstream beneath the Jonesville
bridge.

I bent down to peer beneath the trusses.
I looked, shook my head not believing what I
was seeing, blinked hard, opened my eyes again
and stared.

The canoe was upright, floating downstream as
light as a feather, now that Herb and I and our
gear had all been dumped out.

And in the middle of the canoe, right where I'd
left him, was Satan, sitting on his haunches
and smiling back at us.

How the dog made it through the dunking, I
don't know.

Maybe he did an "Eskimo roll" like they teach
beginning kayakers. Maybe he got dumped
too, but then jumped back in when the canoe
righted itself.

It would sorta be like him to save himself.

I don't know how he did it.

What I saw—and Herb was a witness too—
seemed impossible that day long ago and it
still does.

But it's true.

Despite being dripping wet, with ruined binoculars at the bottom of the river and other gear gone, I couldn't help but smile.

I had to hand it to Satan.

I'd told him to stay, and by golly, he did.

*"He shook his head and claimed to just
be out looking for "sign"—a term hunters
use to cover everything from tracks
to scat. His story was good,
but his behavior wasn't."*

SATAN
WRONG IS RIGHT

CONTRIBUTED BY RICHARD HISLOP

Now, don't get me wrong, Satan did have his good points. In fact, he even helped me on a case once.

I had a tip that a fellow had shot a big doe off Whitney Road in East Fairfield over near the old Pop Austin place. There was no doe season at this time.

Story was the fellow had covered up the doe with plans to pick it up later.

The road was long with few residences along it, which meant if he was lucky and fast, whoever shot the deer might be able to get it out of the woods without anyone seeing him.

Trouble was, I had no description of the fellow or his vehicle. Worse, it was early November rifle season and there were lots of hunters out in the woods. I was busy and had a lot of territory to cover.

161

Wardens can't be everywhere and outlaws know this. Some guys like to play the odds. If they get a tip that a warden is off in another town or the opposite side of a mountain, they will head in the other direction if they are up to no good.

But there's a pretty tight timeline on picking up a downed deer.

You don't want to leave it in the woods more than a day or two tops. Coyotes or other animals might take it and of course, the meat can also spoil.

So, I knew this fellow wouldn't leave it long.

I had Satan with me when I drove on up to the old farm where the caller told me the deer had been shot.

I was pleased to find a nice looking 4WD Nissan pick up truck with gnarly oversized tires and a cap sporting dark tinted windows.

It was parked just off the dirt road with the tailgate nosed into the brush at the edge of the woods and the hood pointed towards the road.

That parking set up was unusual—like someone had the truck all set up to slide a deer onto the tailgate, slap the truck cap shut and high tail it out of the there without anyone seeing exactly what they had loaded.

Very handy.

I parked my truck in front of the truck that looked ready to bolt like a rabbit and then Satan and I went for a walk in the woods.

It didn't take me much of a hike to find the driver. He was slender, in his late 20s, dressed in jeans and a camo shirt with a cap to match —no blaze orange on him at all—with a .30-06 rifle slung over his shoulder.

He was studying the ground so intently that he didn't hear me approach and I spooked him.

He jumped about a foot off the ground when he finally saw me off to the side watching him from about 50 feet away.

"Looking for something?" I asked.

Just in case he didn't recognize the uniform, I introduced myself, asked for his license and asked if he had taken a deer.

He shook his head and claimed to just be out looking for "sign"—a term hunters use to cover everything from tracks to scat.

His story was good, but his behavior wasn't.

He wasn't really dressed for hunting. He had sneakers on his feet. And he kept shifting his weight from foot to foot like he was standing on an ant hill with red ants swarming up over his shoelaces.

He sure acted like he wanted to run. He was from a couple towns over and had a valid license with the buck tag still on it.

He started talking about the deer in the area and his family hunting here for years and all kinds of stories.

And he ended every sentence with a big smile like a used car salesman telling me I was getting the deal of a lifetime on the last purple Pontiac in his lot.

Well, I let him talk and just stood there and nodded, waiting for him to show me where the deer was.

Lots of times guilty guys will glance right over to where they don't want you to look. Or they will make a point of looking everywhere else but where they have hidden something.

They kinda turn their back on that general location or start to wander in the opposite direction, trying to lead you away from what you are looking for.

I just had a feeling if I let this guy keep talking, he would tell me what I needed to know eventually. He was licking his lips between big grins and his eyes kind of rolled around in his head as he told stories to distract me.

And while he talked, Satan did his usual scent and search routine. Not that he was formally trained to do it.

As you may have figured out by now, training Satan to do something was sorta asking too much for a dog with a mind of his own.

He might go along with the concept for a few minutes or he might not. Most of the time—not. I was never sure if he just got bored easily or he just didn't think much of my ideas.

Satan taught me early on that he was happy to make me happy as long as it made him happy too. I was hoping this might be one of those times.

Satan had shown a fondness for fur over feathers.

Getting a deer to jump out of its bed and run

was much more fun than seeing a bird fly away from his perspective.

We'd had several serious talks about this and I had strongly advised him to stick with birds and he'd come a long ways in sticking with that rule. Of course, a leash helped.

But by this time, we'd been together a good decade or so. Satan was slowing down. He was now a "been there, done that" dog.

So I let Mr. Hot Foot chatter on, but I kept an eye on Satan. First and foremost, I was still concerned he might just take off.

But I also thought he might just do the wrong thing—that is—find a downed deer—and in this case, that wrong thing would be the right thing for him to do, you know what I mean?

I could see little beads of sweat start to form on this guy's forehead as Satan trotted in wider and wider circles around us—his nose buried in the beech and oak leaves and then high up in the air sniffing hard.

The fellow was pretending to ignore Satan too. Most people would acknowledge a big black dog running around, make some comment about him or kneel down and call him in and pet him.

But this guy acted like he didn't even see Satan.

That was strange behavior too.

"My uncle took a nice buck off that ledge over there…" the fellow took a step off to the right and pointed, trying to get me to follow his look.

That's when I saw Satan dive in and out of the yellow leaves of beech tree saplings and snake through a thicket of low shrubs off to the left.

I nodded at the fellow and then looked back at Satan. He stopped short and that black tail of his started banging back and forth like a drummer marking the beat for a punk rock band's big hit.

I knew he had found something significant.

He lifted his nose high into the air, straightened his tail like a lightning rod and made a beeline for something off deeper in the woods about 75 feet away.

I marked the spot in my mind where I saw Satan disappear and let this fellow continue to chat me up. There was a big white oak about 10 feet to the right of where Satan was rustling the brush.

I called him back into me. If he had found

a deer, I didn't want him messing with the evidence.

He came back reluctantly, with his head a little low and his tail down too—like he'd done something wrong.

That's when I knew Satan had finally done something right.

I cut the Buick salesman off and said, "Say, let's you and me go see what my dog has found over there."

The fellow's big smile turned upside down and he stopped talking at last.

His shoulders slumped forward if he was a toy and I'd just hit the off switch.

"Why don't you go first? Just head on over to that white oak over there," I said.

I knew by the expression on his face I had the deer and the fellow who shot it in one lucky catch.

He did his best to walk a crooked path over to the deer—like a sidewinder climbing an Arizona sand dune.

I figured he was trying to decide on yet another story to tell me.

But by the time we got to within 10 feet of the carcass, I guess all the scenarios he had been running through in his head fell short.

He sighed, stopped, looked me in the eye and began to tell me the truth. And unlike a car salesman, he didn't have to go check with his manager first either.

"Okay. I guess your dog found it and I'd better tell you the truth," he said as the two of us stood there—make that three—Satan was back with us wagging his tail.

He admitted to shooting the doe, but swore up and down he had seen at least three points on her head before he pulled the trigger the evening before.

Together, we carried the doe out of the woods.

She didn't go into the back of his rig—we loaded her into mine where she would go to feed a needy family in town.

He left the woods with a citation and without his rifle.

Then I opened the door to my truck and Satan jumped in, sat down and smiled big—always happy to go for a ride.

I reached into the glove box where I kept his treats.

Satan knew the routine well and began to dip his head and drool.

I handed him a bone and got to say something to Satan he didn't get to hear often.

"Satan, old man," I said, "you did something right! Good boy! Good boy!"

I patted his head and neck, rubbed his ears and gave him lots of praise.

Then I leaned back in my truck and took a good look at my pooch as he chomped down happily on his treat.

We'd been partners now for a good decade. Satan was a little white around the muzzle now, a little slower but always eager to go out with me on patrol.

Satan and I shared a lot of adventures over the years.

But I have to say, over the course of Satan's long deputy dog career, I came to believe there really was a little bit of devil in this dog.

Coincidence?

Or did I bring this on myself by naming my dog, Satan?

Well, all I can tell you is I learned my lesson.

I named my next dog PB. No, not for peanut butter—for Powerball.

Still waitin' on the big check, but hey...
ya never know!

"Go! GoGoGo!"
one of them is shouting,
"He won't shoot!
He's bluffing!"

Fly Trap

Contributed by Stan Holmquist

Every warden has his favorite locations for staking out night hunters. And in the case of a Rochester landmark I named the Fly Trap, I had 'em all wrapped up in one—my honey hole.

I caught dozens of poachers there over the years.

Why did I call it the Fly Trap?

Well, thankfully, it wasn't due to the bug population.

It was the set up.

The Fly Trap was then—and remains so today —a sweet retreat—maybe 150 acres with a well maintained older home and several outbuildings.

The owners visited now and again, but spent most of their time down country. It had been in the family many years and was a cherished refuge for them.

173

They had a local caretaker mowing the lawn and performing general maintenance. So, the land and home were well kept up.

The house, an old farm barn and shed all rest on top of a gentle rise that leads down to the trees a good 1,000 feet or so away.

And off to the side, sitting a couple hundred feet from the main house but looking over the meadows, is an old school house and maybe 60 feet beyond that a woodshed with a double door —just wide and deep enough to hide my cruiser as the need arose.

Scattered about in the manicured grass leading down to the woods are a dozen good bearing apple trees, annually pruned by the caretaker.

Come late September and early October, the deer flocked to the Fly Trap like fat kids on a diet to an indulgent Grandma. They knew right where the cookie jar was—chocked full of freshly made chocolate chip cookies.

The deer gorged themselves on the apples, putting on pounds in preparation for the long cold winter ahead. Rochester gets deep snows and starvation is a very real threat for the herd.

Best of all from my perspective, was the access to this land—or maybe I should say—the lack of it.

There's only one way in or out.

It's a narrow driveway off a town road and maybe 100 feet before the property opens up on the manicured lawn, there's a metal swing arm gate with big old maples on either side, standing like sentries.

When that gate is closed and locked, ain't no one getting in or out—not driving a car or truck anyway.

It was just a perfect set up for me and my deputies to catch deer jackers.

I had a lot of area to cover and there was just one of me and the occasional deputy. I had to be smarter than the fellows I was chasing.

This was the era of police scanners. A lot of the more serious poachers and outlaws had them. Scanners allow anyone to listen in on warden and state police conversations with dispatchers and between the officers themselves.

The benefit to someone thinking about

committing a crime was their ability to know where officers were working.

If a fellow with an itchy trigger finger and a hankering for some fresh venison knew I was off on the far east side of my district, he figured he'd improve his odds of getting away with shooting a deer out of season by shining and shooting off on the far west side.

So, code names for roads and areas that the dispatchers and state police understood, but the outlaws didn't, was the way to go.

My Fly Trap was named after the Venus Fly Trap, a plant native to South America.

This plant lures in flies with a bit of sweet nectar and then clamps its jaws shut, traps them and gobbles them up—no escape.

The Fly Trap's owners were more than happy to have me routinely patrol here.

I worked with the caretaker to make certain the gate hinge was well greased so it wouldn't squeak when we closed it behind any car or truck that rolled in.

And I'd coordinate with the owners to make sure

the woodshed's doors were unlocked too when
I wanted to hide inside.

It was a sweet set up—just like that South
American plant.

Of course, unlike flies, poachers generally don't
come alone. They often have two, even three or
more guys riding with them.

So, most always I had a deputy along with me
when doing surveillance like this.

For one thing, I needed someone to shut the
gate while I followed the car or truck on foot.

But more importantly, it was a lot safer that
way.

If a warden confronts three or four guys
in the middle of the night in the middle of
nowhere it can turn into trouble.

Often, these guys have been drinking and aren't
thinking straight. Some get mouthy. And of
course, they are carrying loaded weapons.

So, I had to be smart about how we went about
approaching any car or truck.

On this particular night, I was with my deputy,

Henry Giddings of Pittsfield. Henry is tall, fit, handsome and always carried big revolvers with him. He had one that I swear had a barrel that reached down to his kneecap when he holstered it.

I used to joke it would take a weaker man two hands just to pick it up, let alone hold it steady, aim and fire.

I often wondered how many hours Henry spent actually shooting the thing. The pistol couldn't really be fun to shoot because it is so heavy.

But it sure impressed the scofflaws we caught. When Henry trained the sights of his Dirty Harry special on 'em, it was like they were looking down the barrel of a cannon.

Guys who stepped out of a vehicle with an attitude lost it fast when they saw the armaments Henry carried.

I was comfortable with him carrying super sized shooters. I knew Henry was sensible and he would never pull the trigger unless we were up against it.

In fact, Henry did an exceptional job over the years. He never failed to back me up and he also had an effective way of calming me down when I needed it.

I don't think anyone ever noticed it either.

Henry's technique was to slide me a Necco wafer —those little nickel sized hard candies that came in different flavors and were sold in rolls. I don't know if they even make 'em anymore.

Henry loved 'em and it seemed whenever we were on patrol, he had a roll of them in his pocket and sucked on one or more during a patrol.

One time in particular I remember we had a couple fellows who thought pretty highly of themselves and didn't have much regard for the law.

We'd pulled them over on a back road in the middle of the night for shining a field.

The two of them came charging out of their car cursing and waving their arms, calling me every name in the book and running for my cruiser.

Situations like this can turn bad real fast.

You don't want to escalate things by getting into a shouting match and at the same time, you can't be a doormat.

You gotta take charge.

And generally speaking, you get just one shot to
take control.

In this case, I told the driver—a big man who
was charging at me like a bull—to "STOP!"

He stumbled but recovered, stood back up,
lunged forward again and kept up the epithets.

There wasn't even time to pull my revolver. He
was right up on me and twice my size.

I stepped aside, reached out and grabbed
his shoulder and spun him away from me.
Then I pushed him hard from behind with
both hands.

There was a telephone pole next to the road and
I steered him towards it.

He face planted right into it.

He stopped dead, a shudder running through
him. I swear he shook the pole a little too.

I hoped the smell of creosote and the possibility
of splinters the size of pencils poking his face
like porcupine quills would convince him to
calm down.

Meanwhile, Henry was handling the passenger

—a smaller lad who had wanted to back up his big and very angry buddy.

This youngster had come out of the car with big ideas too, but once he cleared the taillights and saw Henry had drawn his Dirty Harry revolver and was standing in front of him with a finger on the trigger, well, this youngster stopped dead in his tracks.

In fact, he stopped so short he slid on the loose stones at the side of the road, his feet went out from under him and he landed flat on his arse —his mouth wide open and speechless.

Henry had that fellow well under control.

I was some put out with the very idea that these two fellows would even think of threatening two lawmen like this and using such foul language.

So, I kicked the feet of the pole leaner farther apart like I was getting a Morgan show horse to park out.

I told him to put his hands behind his back.

I was going to cuff him, check him for weapons —lots of guys carry a pistol in their pocket or a hunting knife the size of a dagger under their coats or in a boot or both—and take him in.

I was steaming.

Henry was talking to the other fellow, telling him to put his hands behind his back and stay put.

That fellow was now quiet as a lamb and saying, "Yes, sir," to Henry. The fight was out of him like a popped balloon.

Henry came on up behind me to see if I needed help. I said, "Give me the cuffs, Henry."

Well, like I said, Henry knew me well. We'd been on patrol many years together.

He could tell by the tone of my voice that I was a quarter inch away from rocking this guy's world so he would never ever think of calling a lawman names like he had just called us ever again.

But the fact was, this bruiser had already changed his attitude.

Maybe having a little quality time with his stubbly chin planted on that stinky creosoted telephone pole reminded him of a time out when he was a kid.

I don't know.

But in less than five minutes the big man's tone had gone from bar room brawler to first grade boy whispering his request to the teacher that he had an urgent need for a hall pass to the boy's bathroom.

Henry had the benefit of standing back from the situation a good 15 feet.

He shined his flashlight onto the fellow and saw a new look on his face—fear.

I was still seeing red, my teeth clenched, my hand out behind me waiting for Henry to slap the cuffs in my palm.

Adrenalin filled and focused me like a laser.

I waited with my teeth clenched.

That's when I felt something small and chalky, the size of a washer pushed into my palm instead.

It was a single Necco wafer.

The cuffs came five seconds later—laid right on top of the candy.

That featherweight bit snapped me out of my desire to teach this fellow a lesson.

Henry did that fellow a big favor that night—
though the guy never knew it.

I took a step back, then another, stood up
straight and took a deep breath—getting some
perspective on the situation.

I saw the big man's shoulders heave up and
down and fat beads of sweat drizzle down his
head despite the frost in the night air.

Suddenly I was able to see him as just another
boozed up local boy who had made a mistake.

In fact, he was now begging me to let him stand
up straight and using words like "please" and
"sir."

"You going to behave?" I asked him.

"Yes, Warden. I'm sorry. I'm sorry," he said in
a slurred but sincere apology.

One glance at the other fellow and I knew from
the look on his face he'd be telling all the big
guy's pals about how fast he went from brawler
to bawler.

That would punish him for the next 40 years or
so.

Yup, Henry saved my bacon a bunch of times.
He and the other deputies who worked alongside
me were all good men.

But I digress.

Back to my Fly Trap adventure.

It was a lovely September evening, a
moonless night and Henry and I were at the
Fly Trap, having parked the cruiser inside the
woodshed where any night visitors would
never see us.

Then we went up and sat down on the side hill,
watched and waited.

Around 11 pm, we saw headlights coming slowly
down the private road.

Henry knew what to do. He grinned, nodded at
me and without a word he scooted back along
the tree line in the dark to the gate.

He passed right by them; no more than 30 feet
away and they never saw him. The driver and
his crew all had tunnel vision, staring straight
ahead.

I knew I could count on Henry to quietly drive

the cruiser out of the woodshed, lights off, shut and lock the gate and wait for me.

I walked down the slope and fell in step behind our visitors' car. I was no more than 15 feet off their bumper.

Funny thing about poachers—they are so eager to shine a light and take deer they almost never look in their rear view mirror.

If they did, some of them would have a heart attack.

There were many times I was walking just a few feet off a paocher's rear bumper. And that is what I did here.

As soon as I had written the license plate number and the make and model and color of their car—an early 1970s metallic green Dodge Coronet that had seen better days—I scooted over to the hillside behind and above them to watch what happened next.

I knew from experience where the poachers were most likely to be stopping and taking aim—on the slight slope overlooking that sweet little apple orchard.

When I arrived, the crew was there all right.

Headlights were off—just parking lights on now.

They were setting up, with a shooter leaning out the passenger window and a fellow in the back seat shining the meadow for him with a bright light.

There on the grass, I had a front row bleacher seat watching the whole thing unfold.

The eyes of the deer were reflected back to me as well as to them. There were four or more deer there. Two of the sets of eyes were lower than the others—likely this summer's fawns.

When someone talks about "standing there like a deer in the headlights" it's very true. Deer do in fact stand still as a statue when you shine a light in their eyes in the middle of the night.

To give the animals some credit, deer aren't used to artificial light and don't generally carry flashlights with them. So a bright beam trained on them is both blinding and an entirely new experience.

They are just trying to see and figure out what is going on. Fact is, a lot of people do the same thing—freeze—when you shine a big spotlight on them.

The thing is, while the deer stands there trying to figure out what is happening it generally gives the cheater with the rifle in his hand enough time to lean out the car window, point, shoot and down the animal.

That's the way it is supposed to happen in the Perfect Poacher's Play Book anyway.

But sometimes, it doesn't. Not surprising.

If poachers were really any good at hunting, they wouldn't have to cheat, right? So it kinda makes sense really that even when they give themselves a huge advantage, sometimes they still screw it up.

And for these boys, it was looking like one of those nights.

The light was shining bright. A shot was fired—a flash of gunfire came out of the rifle muzzle, and echoes bounced off the surrounding hills.

I looked.

No deer went down. But the herd did run off.

The fellow with the rifle had just flunked Poaching 101 in a car full of his peers.

Hunh. The marksman in me wondered whether it was the shooter, the rifle, nerves or all of the above.

For sure, the guy couldn't argue the sun got in his eyes or the deer moved.

"I don't see a deer! I don't see the deer! Do you see a deer?" was the first yelp.

That was followed by, "Are you kidding me? It was standing right there!!!!" by a new voice.

"You idiot!" and a scramble of voices and insults and accusations rose up into a big male wail.

Each one accused the other of contributing to the failure to down a deer.

The cursing and hollering, slamming and banging that flowed outta the window of that bucket of rust was something to behold.

The car rocked back and forth and side to side as they all literally jumped out of their seats with rage and accusations and excuses.

I just sat back and watched.

There was no need to rush down and nab 'em. I knew Henry had shut the gate.

I decided to just see how this played out.

I stood up on the hill smiling despite myself and waited to see what happened next.

Would they shine a second field and take another shot or would they bolt?

After three minutes or so the brouhaha inside the four door beater seemed to be winding down. I saw the dome light come on. The driver and front seat passenger opened their doors, stepped out and changed places.

Interesting.

I might be about to get two shooters with one stake out if I just sat tight a little longer and watched.

In less than a minute, the new driver was rolling the car up the lane and the interior quieted.

My bet was the new driver was headed 700 feet or so around the bend of trees, to the second meadow on this farm—hoping there were deer feeding there and the earlier shot had not spooked them.

I got my feet moving and trotted along the grassy hillside slightly behind and above the car.

Experience told me they wouldn't think to look back unless they saw headlights following.

Their car rolled slowly past the boarded up summer home, swung right around the bend past the sugar maples.

Sure enough, when the driver came to the second meadow, he slowed and stopped.

The spotlight came alive again from the back seat. The shiner moved the beam slowly from right to left.

I stood off behind them and watched.

I saw deer about 100 yards away.

I was less than 40 feet from the poachers' car now.

These fellows were so intent on their spotlighted quarry they were only peering at whatever was inside that white beam.

They had no idea a warden was practically in the car with them.

I was close enough to catch a lot of their conversation. These guys were as squirrely as fifth grade boys at their first school dance talking about the girls.

When the light shone on a group of three deer
—one of them a nice sized buck—it was like the
prettiest girl in school had just smiled at them.

"Oooh! Ahh! Man! Do you see that one???"
I could make out four different voices coming
from inside the car—which rocked side to side
from their jumping around inside the old tin can
trying to get a better look at the deer.

"Shut up! Shut up and hold still!" I heard a
voice hiss in a whisper that was as loud as a
shout.

It must be the fellow with the rifle trying to aim.
I saw the rifle barrel come out the passenger
window again and reach long.

The light was still on the deer.

But if the light didn't spook them, their crabbing
and jostling did.

I saw one set of eyes start to move off and then
another pair turns away. The deer were headed
out.

"Shoot! Shoot!" someone yelled.

BOOM!

Shot fired.

I leaned in and squinted a little, fully expecting to see a pair of eyes lower and a deer fall mortally injured.

The fellow in the back seat with the spotlight did too. He was going wild with the beam—winging it right to left, up and down like it was strapped to a bull rider's free hand on a championship ride aboard Bodacious.

I saw the eyes of two raccoons high up in a tree for a split second the guy was so wild with his light.

I mean what were the odds two guys had both missed deer this night?

There had to be a deer downed this time, right?

It took a few seconds for this shooter, the passengers and even for me to have it register in our brains, but the would be poachers—who were essentially shooting fish in a barrel—had missed yet again.

Different rifleman, same outcome.

There was silence in the old Dodge for maybe

three or four seconds and then, it was if someone had found a rattlesnake under the seat.

All hell broke loose.

"Where is it?"

"Shine the light! Shine the light RIGHT!!"

"Are you KIDDING me?????"

It was like the prettiest girl in the school had come on up to the smitten fifth grade boy to ask him to dance and he threw up on her party dress.

It was that bad.

The kind of male mistake your drinking buddies never let you forget.

One of those tales they will tell about you at your funeral in hushed tones and laugh like crazy—no matter if you lived long enough to find Jesus, rescue a family of five from a burning building or give a million dollars to the noblest cause in the world.

To these pals, you are forever a moron and they ain't never gonna let you forget it—not even when you are six feet under.

Nope, this guy would never live this one down.

I actually felt a little bad for the guy.

This was the kind of screw up that prompts men to quit their jobs, leave the state and change their names rather than put up with the teasing that is certain to dog them for the rest of their lives.

If there was a winner so far tonight, it was shooter number one. Suddenly, he wasn't alone in his incompetence.

The old, "SEE?? It's not me, it's the gun!" argument.

And in fact, that's the defense I heard after more groans and cursing and accusations came rolling out the windows.

Once the Coronet chorus determined their second sniper had missed by a mile as well, the noise inside the old Dodge reached a whole new level of outrage.

"I told you! I told you! It wasn't my fault I didn't get the first deer!"

Any thought of being quiet so as not to scare the

herd vanished along with their visions of fresh venison.

They were calling each other names and yelling as if they were jumping on and off bar stools at the Eagle Tavern watching the Yankees play the Red Sox with money bet on the game.

The argument went on for a good minute or two before one of the fellows in the back finally came to his senses.

"Hey, come on! Stop fighting! We gotta get outta here," I heard him yelp a time or two.

He must be the designated brains of this outfit, I thought to myself.

As a warden, it helps to know there is at least one guy in a bunch of outlaws that has some sense.

"Dave's right!" another voice piped up. "Let's get outta here. Someone might be coming. Drive!" another voice said.

It was time for me to catch up with Henry at the gate and for the two of us to wait for them to roll on up.

Their bad night was about to get worse.

Their car had a long gravel loop to traverse before it reached the gate. I had a nice trail through the woods—and I used it to head them off.

When I saw their parking lights rounding the bend, Henry and I were ready.

He'd closed the gate and was standing off to the side ready to back me up.

The cruiser was a little off to the side, ready in case we needed it.

I was standing 25 feet or so inside the Fly Trap gate with my flashlight in my left hand. My flashlight was off and my right hand was on the pistol grip of my .357 revolver, ready to draw it if I needed to.

The old Coronet was coming on slowly like a boat riding two foot swells.

The driver was weaving around some potholes, plowing through the late fall mud and coming right on to me but had not yet seen me, Henry or the cruiser.

I raised my flashlight, hit the switch and shined it right at the windshield.

Time to see what these boys do when a bright

light is shined in their eyes. Would they freeze or would they drive right over Henry and me?

Most guys, even those up to no good, when they see the big man in the hat with the badge appear outta nowhere in the middle of the night, training a light right in their eyes and yelling, "Game warden! Stop!" do what they're told.

It's the sensible thing to do.

Not this guy. He keeps coming. He was being egged on by at least one of the guys inside.

Their windows were down and I could hear them.

"Go! GoGoGo!" one of them is shouting, "He won't shoot! He's bluffing!"

I pull my .357 and point it at the windshield and shout again, "Game warden. Stop!"

The driver weaves the front of the Coronet off to my right, like a soccer player considering a corner kick on the goal.

The car is rocking and rolling like the Shakin' Aiken ferry running between Essex, NY and

Charlotte in a stiff north wind.

I could actually hear the leaf springs squeak as the men threw insults and accusations at one another and fought over whether they should drive through the gate or stop.

I heard the "gogogogo" guy say, "We can make it! They only shoot your tires! GO!"

I don't know what bad advice the fellow had gotten from his grandpa, but the shootin' out tires chapter of warden enforcement had ended two or three decades earlier.

When faced with deadly force—like a Dodge Coronet carrying four men and one or more loaded weapons bearing down on top of you—we were authorized to shoot the outlaws.

I mean think about it, it wasn't the tires causing the problem, right?

I didn't have time to explain the change in procedure to the Coronet driver or his delusional and misinformed friend.

I raise my weapon and point it at the windshield directly at the driver's seat.

Henry pulled out his cannon.

"Stop or I will shoot YOU," I shouted, trying to add a little clarity to what anyone with any common sense would have figured out some time ago.

"He's bluffing! He's bluffing! GOGOGOGOGO!" the same voice is still shouting, urging his buddy on and waving his arm out the passenger window like it would make the car go faster.

I took a bead on the windshield. I was a count of two from firing at the driver.

That's when I heard a voice inside the old Dodge shout, "Shut Up!" followed by a dull thump like a pumpkin thrown against a wall.

The gogogogogo guy's yelps were suddenly nonononono more and I saw the arm that had been flailing outside the front passenger window wilt like a piece of over cooked asparagus and stick there.

Someone inside had hit him hard and shut him up at last.

"Someone in this car has some sense," I say to myself.

The Coronet driver slams on the brakes. The car's nose dived into a mud puddle and its square behind slides a bit sideways before it comes to a stop.

No doubt the gang had taken everything out of the trunk earlier in the day to make room for all the deer they intended to put in there this night.

I keep my revolver pointed square at the driver and begin walking in towards the steering wheel, while Henry covers the other side.

I shout to the driver to turn off the car and throw the keys out onto the grass. I had them get out and keep their hands up.

When the occupants get a good look at Henry and the cannon Henry's got pointed at them, they go as silent as worshippers at a Sunday church service.

If any of them had thoughts of putting up a fight, that idea was gonegonegonegone after seeing Dirty Harry.

I had them put their hands on the hood.

It didn't take me long to figure out who the big mouth with a death wish was.

I couldn't help but notice a skinny fellow in his

late twenties wearing paint stained Dickies, sporting yellow teeth and a red bruise on his forehead that looked to be growing into a big egg.

They were all pretty quiet, until the driver—a pale kid maybe 18 years old—asked me, "Is it true Warden? Would you have just shot out a tire?"

"Son, you were a count of two from me shooting you right in the head," I said, looking him straight in the eye.

I paused and added, "And unlike you guys, I hit what I shoot at."

I looked over to see if my comments had any impact on these fellows.

The egghead lowered his gaze and finally stopped fidgeting.

The other two looked down and shook their heads like they couldn't believe whatever they'd been drinking earlier that night had led them into this ill fated escapade.

I turned back to look at the driver.

He wobbled, his face went from pale and freckly to ash gray. His eyelids fluttered a couple times, his eyes rolled back into his head and then his knees buckled.

He collapsed like wet laundry and fell onto the grass.

I shook my head in disbelief and then turned to the egghead and his two buddies.

"Well, Congratulations, Gentlemen!" I said. "You boys finally managed to knock something down tonight."

*"He was so hung up in the wire,
he looked like he was levitating
above the ground on his back—
a middle of the night,
middle of the woods séance."*

TRACK STAR

CONTRIBUTED BY DENNY GAIOTTI

I t was a late Saturday summer night when the call came into my cruiser about a dead deer over off of Route 116 in Middlebury.

The dispatcher told me the deer was off Cobble Road, near a turn out on the west side of Route 116, near a spring.

That was the opposite of where the hood of my car was pointing, of course.

I pursed my lips and sighed just a bit and nodded to the dash as if the caller was sitting there talking to me.

If folks would just slow down.....

"Any driver on the scene or a disabled vehicle?" I asked.

"Negative," she said.

Well, it was bad news for the deer, but some

local family in need of food would likely benefit.

I knew the area well—a lot of deer liked to travel through there.

But folks tend to drive a little fast on that stretch—not a lot of places in town where you could burn the carbon off your spark plugs, as they used to talk about in the days before fuel injected engines.

This was one of the few—and kids borrowing dad's car on a Friday or Saturday night really loved this road for that reason.

Burning rubber and a bit of drag racing was not uncommon.

"10-4," I told the dispatcher. "I'll head over."

I slowed the cruiser, looked in the rear view mirror, saw no one was behind me, headed for the ditch and spun the steering wheel in a big 360.

Just before I ditched a wheel, I punched the gas.

The old Plymouth Grand Fury took up most of the road. It was kind of an art to be able to spin 'em like this—without having to do a three point turn.

Those of us who'd worn the badge awhile took pride in being able to reverse direction like a pro.

It was a small thing, but it was one of the skills that separated the men from the boys.

I wasn't rushing over intending to attempt to revive the deer. It's all about preventing another accident at this stage—and if the downed deer was a doe, maybe locating an orphaned fawn or two.

A car hitting the carcass could cause a serious accident and yes, could even get someone killed.

It took me a little over 15 minutes to get over to the other side of town.

Trouble was there wasn't a deer in the road.

I called back in to the dispatcher and asked if there were any more details or a more precise location.

I heard rifling through some papers come over the microphone—this was all before computer screens. The only sound you might hear is finger tapping on typewriter keys.

Good penmanship was one of the requirements for a Dispatcher's job.

"Negative, just 116 East. Turn out on west side, near spring. Nothing more."

"10-4," I said. "I'll get out and take a look."

There were a number of possibilities as to why I couldn't find a deer.

It could be the caller was wrong—the deer wasn't dead, just down and stunned and had gotten back on its feet and run away.

It could be it was a prank and there never was a deer hit.

It could also be that someone came along and grabbed the carcass and it was already hanging in a garage and about to go into some local's freezer.

It could be someone called it in so they could be off somewhere else poaching on another side of town and they wanted me as far away as they could get me while they were off shining a light and shooting in another field.

All distinct possibilities.

I pulled off the road onto the grass, turned off the cruiser, toggled my parking lights on, grabbed my flashlight and headed out into the night air for a closer look.

There was only the slightest edge of a moon hanging in the sky, not enough to help me.

Shining my light side to side across the asphalt as I walked, there were no fresh skid marks, no smell of rubber in their air or burning brakes, no obvious blood or hair in the road.

But I wasn't going to give up that easily.

I went beyond the intersecting roads thinking maybe the caller was just giving the dispatcher a rough reference point.

I walked looking for any sign that a deer had been hit—a piece of plastic from a broken parking light, deer tracks in the soft roadside dirt, a tuft of brown or white fur, a patch of tall grass knocked flat at the edge of the road, engine oil or antifreeze drips from a vehicle on the road—anything.

I'd gone maybe 100 yards when I heard a rustling off to the right—maybe 20 feet off the road in tall grass and cattails.

I stepped up my pace and headed for the sound.

A pick up truck pulled up behind me on the road and kinda swerved away towards the opposite shoulder as he got closer. I turned to see a twenty something driver gawking at me from behind the windshield of a seen better days dark green Ford Ranger.

He had a look of surprise on his face—like he didn't expect to see me here.

His window was down, but he didn't wave or lean out his driver's side window to say hello or stop to offer any help or even ask what I was looking for at 1 am.

He rolled on past and then he stepped on the gas hard once he got past me and sped off.

That behavior was not your typical friendly Green Mountain resident late night reaction to an officer walking down a country road in the middle of the night.

Most folks will stop and ask you what's going on and maybe even offer some assistance.

I focused on the Ford's license plate as he took off. It was green with white lettering and pretty beat up—like an alligator had chewed on it—and the tailgate—which was down like it was ready to load something—didn't look much better.

I made out at least three numbers on the plate —a 4, 8 and a 3. I repeated them over and over to myself, to memorize them just in case I might need to find that driver later. As I repeat the numbers to myself, I walk and shine my light from side to side and peer into the grass alongside both sides of the highway looking for the deer.

I didn't want to change my pattern just in case the fellow in the Ranger was looking back at me through his rear view mirror. I didn't want to spook him.

I had a sneaking feeling he and I might be looking for the same deer.

Why else would he be out here alone at 1 am?

I walked to within 15 feet of the spot where I'd heard the brush rustle when a twig snapped in the same location.

Whatever it was up there had some heft to it.

I stopped and shone my light dead on the sound —back and forth—over the tall grass and into the puckerbrush and cattails beyond.

That was all it took.

A skinny fellow jumped up straight as an arrow, like like he'd stepped right on a yellow jacket hive. His arms were at his sides. He stared right at me, his eyes and mouth wide open, looking right into my light.

Then he spun away from me, bent over and took off like a Sasquatch launching from an invisible weedy starting block—a light coat billowing out behind him.

"Game warden. Stop right there!" I yelled.

Well, apparently this baby Bigfoot had a good bit of Hotfoot breeding tossed in or he didn't understand English.

He was headed through the old farm pasture for the tree line. I trained my light on his back and watched him fly.

He looked to be wearing a shiny dark blue windbreaker and jeans.

He had good form—his arms were pumping in rhythm, his knees snapped tall, his head was up and back a bit—but not so's I could get a good look at his face.

Trouble was he headed towards a sloppy track.

I took a split second to dig a boot heel down and twist it hard into the loose gravel to mark the spot along the road, before I jumped over the bank.

I wanted to see why he'd been so eager to run in the first place.

It might be kinda foolish, but it wasn't illegal to be hiding in the grass beside the road.

Sure seemed like he was up to something.

I grabbed a fistful of the grass and twigs with my right hand and bent them over good too, just in case I needed another way to find this spot later.

There weren't many reasons for a fellow to be out here in the middle of the night crouched in the brush. If he was a vagrant just passing through, there were a lot better places to camp for the night.

Experience told me he might just be looking to fill his freezer with venison without the trouble of hunting.

Sure enough, right where the fellow had blown

out of the bush, there was a dead deer—a nice buck.

I shone the light over the body quickly.

There was no obvious gunshot wound. Looked like a vehicle had hit it.

The caller got the dead deer part right. But the "lying in the road" part?

The buck had been moved and hidden.

Lying next to the carcass was a knife with a clean blade and a flashlight. I guessed the shadow that had lit off for the woods was about to get down to business when I interrupted him.

Well, that's not the way we do things in Vermont.

You don't just get to pick up dead deer and take 'em home and turn them into freezer fillers like you won the road kill venison lottery without placing a call to a warden to discuss the matter.

All this looking at the scene had taken me about 10 seconds. It was time to launch the rocket boosters and get after this fellow.

I sighed, hoisted up my belt, took a deep breath and stepped out big.

The good news was the runner couldn't have
done me any more favors if he sent up flares
and tied ribbons to the cattails ahead of us. Wet
ground ahead.

From the brief glimpse I got of him before he
blew up a like a partridge with his blue coat
flying behind him like a sail, he looked to be a
skinny guy.

I was hoping he was dumb enough to smoke a
few packs a day. That would cut his wind and
slow him down substantially.

The race was on.

Unless I found a wallet or a hat or something
I might be able to trace back to this darting
shadow, if he outran me, it was pretty
unlikely I was going to be able to build
a case.

I could think of a good number of suspects, but
most didn't have the speed this fellow did. And
not many had the audacity to ignore a direct
order to stop.

I dug in and ran.

Well, to be technically correct, due to the terrain,
I hopped and skipped.

There were dozens of marsh humps, cattails and standing water for a big man like me to bust an ankle on. The only good news was that this wet stuff sure made the tracking easy.

I saw by the footprints around the carcass this fellow was probably wearing some kind of canvas sided, wimpy soled sneakers.

That brought a little smile to my face as I launched after him.

When he lost one or both of those flimsy fashionable foot wear in the famous Addison County blue clay mud and maybe had his socks pulled off to boot, he might just reconsider turning around and talking to me.

In fact, if he stuck it out that long, he might even ask me to carry him out.

Addison County clay is famous for it's boot sucking blue clay. The stuff sticks to your feet like cement and is five times as slippery.

So, to get across this swamp it was critical to keep moving.

If you stop and think about your next step, you are going to sink up to your knees in a matter of seconds and be begging for a helicopter ride.

I shone my light on across the wet pasture and saw him a good 150 feet ahead of me, walking as if he was on a high wire in a big wind—his arms out to his sides, battling the bog and looking like he could fall into the muck at any time.

Even though I had the disadvantage of being a big man on soggy ground, I had the advantage of my big boots working like snowshoes—they helped keep me on top.

I knew I could track him, but catching him might be another matter if he got too much of a lead on me.

I was hoping he'd fall flat on his face in the muck and that I wouldn't.

But either way, I couldn't just let him run without a chase.

It took a good bit of dancing but I made it to the tree line and drier ground. I could see his tracks when I entered the woods.

I bent down and sucked in a couple of deep breaths, stood up and shone my light around into the forest edge trying to pick up his trail.

The trees had lost a lot of their leaves thanks to an early frost. As I arced the beam through the trunks, a big splash of blue caught my eye. I headed for it.

It was a nice blue nylon wind breaker—half of it wrapped up and wound tight into a rusted barbed wire fence.

The color looked an awful lot like the one my runner had been wearing 10 minutes earlier.

I grinned.

Looks like this fence had slowed him down some.

The leaves were all kicked up and the ground was stomped, mashed, kicked, beaten and scratched up looked like he'd been fighting at least two other guys and maybe a tiger.

Soft-soled sneaker tread marks spread within a good four foot radius.

By the looks of the dust up, I guessed he'd been hung up a good couple minutes.

That woulda slowed him down, shook him up and made him think twice about steaming full tilt through the woods in the dark.

All in my favor.

Good old barbed wire. The bane of beasts and bumblers, but a boon to them that wants to bring 'em up short.

I smiled and unwound the light weight jacket from the fence and stuffed it inside my coat to keep as evidence.

My runner must have heard me coming or seen my torch shining and just wriggled right out of his jacket rather than take the time to free himself and take his coat with him.

That meant he was scared, tired and likely to make more stupid decisions.

All good.

Now that I was on solid ground, I shut off my light and opened my ears to the night.

If this fellow didn't know this piece of Middlebury, he was about to run into more surprises. He'd run into more cattle fence and a slippery steep shale ledge soon.

It would make for darn tough going on a dark night without a flashlight.

I listened for the sound of a guy smashing through the woods—the snap of paper dry pine branches, leaves getting scuffed up, wheezing, coughing or even the sound of water.

I'd hiked this area before. I knew there was a creek up ahead feeding this wet pasture. If the runner hit that unknowingly he'd make quite a splash as well.

But what I did hear, I didn't expect at all.

"Helllllllp."

It was like a little girl's voice up ahead in the dark.

It wasn't a manly man yelp of a guy drowning or needing to get yanked out of a burning building or sinking into quicksand.

It was the cry of a guy almost too embarrassed to be asking for assistance.

I stood there wondering if I was imagining things.

I cocked my head like an owl, leaned forward and listened harder.

Ten seconds later, I heard it again, "Helllllllp," a little louder this time and with more anger and urgency in the tone.

I got a general bead on where the call was coming from.

It was maybe 200 feet ahead of me and off to the right by another 75 feet or so.

I had no idea if the yelper was the fellow I'd been chasing for a good half hour, but whoever it was, I had to check it out.

I turned my flashlight back on and passed it to my left hand. I unsnapped the leather keeper holding my revolver tight in its holster.

If this was some sort of trap, I wanted to be ready. I walked in with my torch held off to my left, away from my body and my other hand on the butt of my pistol.

As I got closer, I heard a new sound—twanging. Like someone restringing their steel string guitar and doing a really bad job of it—the sound was up and down the scales willy nilly.

Along with the twang was some tin can kinda rattling followed by scuffing feet, shuffling, huffing and puffing and a goodly stream of colorful cursing.

I parked myself behind the biggest tree I could

find—a nice old white pine—and peered hard into the dark.

What in the world?

If my runner was gonna get himself stuck, that shoulda happened back in the bog dance a half hour earlier.

But there he was.

All trussed up slicker than if Houdini's helpers had tied him.

My suspect was suspended above the ground, his arms tight to his sides, on his back looking up into the night sky through a thick canopy of old apple trees and red pine. He had on a faded and outdated rock band hero Tshirt and his pants had fallen so low his tidy whities were barely covering his pale behind.

He was so hung up in the wire, he looked like he was levitating above the ground on his back—a middle of the night, middle of the woods séance.

Kinda spooky actually.

I watched him for a few seconds—long enough to see there was no one else around, he didn't have a weapon and he really was stuck.

I saw his chest rise preparing to suck in air
to make another blatting yelp for help, hit my
flashlight beam and trained it right on him.

He jumped like he'd been stung by a hornet
and let out a frightened yelp. He was an angry
papoose stuck in a steel hammock.

The old fence twanged and boinged like bad bed
springs under a fat man, but held him fast.

There was no need to say, "Freeze!" or "Halt."
He wasn't going anywhere.

I couldn't help but bust out in a smile at his
predicament.

I walked on over and saw how his levitation trick
was done. He was all twisted up inside three or
four strands of rusty barbed wire. The old fence
was hidden behind coiling grape vine, burdock,
thistle and tall grass.

The youngster had run full tilt and slammed
right into the lines, flipped himself over once or
maybe even twice. When he struggled to free
himself, he just drove the barbs deeper into his
jeans and Tshirt.

He was trussed up tighter than Auntie Beulah's
Thanksgiving turkey.

223

Seeing him like that, I had a mix of emotions running through me.

First, I was happy my foot race was done.

Second, this was some funny predicament the fellow had gotten himself into. It didn't look like he was really hurt, just nicked up a bit. But he sure was stuck.

Third, sure wish I'd put a camera in my pocket. This was one for the Poachers Hall of Shame for sure.

I took a step towards him and he yelled again, louder, "HELLLLLP!"

I guess he thought I might walk away and leave him.

Well, it would serve him right to spend the night in a barbed wire jail, but I put my anger aside and walked in towards him.

"Hold on, I'm coming," I said. I lowered my light and walked in the final 50 feet.

I parted the brush on either side of him, and looked close to determine the best way to extract him from his cage.

I thought for a minute I might have to flip him
like a pancake on a griddle to untangle his hide.

I wondered if anyone made a human sized
spatula for such occasions.

He looked at my face like a four year old who'd
taken a tumble off the monkey bars, skinned
his knees and was too afraid to look and wanted
Mom to fix it.

I did a couple "Hmmmms" and "Unh-huhs" like
a doctor over a nervous patient.

Then I nodded, stepped back a couple steps and
lifted my light until it was hitting him full in the
face.

"I can get you out. But how about first you tell
me what you were doing with that deer?"
I asked.

Knowing I had the power to free him, the
youngster got pretty chatty.

"I was driving home from the Alibi and found
it on the road. It was already dead—honest. I
drove over to my buddy's. He's got a pick up
and lives close by. He dropped me off with a
knife and said to dress it out and he would
come back and pick me up,"

He musta felt another stab on his backside. He paused, wriggled and bounced the wire a couple inches and winced, then added, "You showed up and I ran."

"Hey, get me out of here, will you, Warden? Please?"

What he said made sense.

I was thinking back to the pick up truck that rolled slowly past me and then sped off. That had to be his friend coming by to pick up the deer."

"Your pal drive a Ford Ranger?"

His head jerked back and his eyes opened wide.

I surprised him with that one.

"Yeah. How'd you know that?" he asked.

I wasn't about to answer his questions.

"Put your feet flat on the ground and hold 'em still," I said. "I'm going to spread the wire. See if you can wiggle out—limbo like. Watch your eyes. If this old fence breaks, you could lose one."

He did as he was told and with me lifting up on the wire at his chest with two hands and pressing down on the strands beneath his behind with a size 12 boot, he managed to wriggle out.

It took awhile. He moved slowly and there was a goodly selection of "ouch" "awww" "geez", and "OW!"s along with a few words I won't repeat.

When he got to his feet, I saw he was a good six footer, slim build. Brown hair and eyes, still fighting some acne. I guessed he was not long out of high school.

I looked at his feet and saw his sneakers were still on, but covered in Addison County's finest blue clay up past his ankles. Small wonder he couldn't get his feet underneath him to push himself out of the wire strands—it was like his soles were slicked with an inch of Bag Balm.

He started to inspect his hide, reaching around and under his torn up T feeling for scratches and nicks.

I interrupted him.

"How about your name and some ID—like your driver's license," I said.

"Okay," he said, reaching into his back
pocket—which was about down to his
knees.

"But could you help me find my glasses? They
flew off when I hit the fence here. I can't see
anything without them."

I took his wallet, opened it up to see his driver's
license, looked back at him and said, "You stay
right here."

He nodded and said, "I'm done runnin'."

As I walked around looking for his glasses,
I saw him continue to pat himself down,
looking to see how many holes he'd punched
into his hide from his run through the woods
and his barbed wire bash.

I found his spectacles maybe 10 feet from
where he landed in the fence—they were
sitting cockeyed on top of a thick tangle of
honeysuckle.

The silver frames glowed in my light.

I brought them back to him and said, "You're
lucky. They don't even look broke."

He put them on and I saw a smile of relief cross

his face. He threw his shoulders back and knocked his heels together to kick some of the clay off his sneakers.

Now that he was extricated from his prison and could see again, his attitude began to change.

The 20 something testosterone fueled attitude returned.

"You want your jacket?" I asked him.

He took it and said, "Yuh. Thanks."

Then he looked me over—a guy twice his age and more than twice his weight—and said, "You know I was a track star in high school—Otter Valley."

I was copying his information down from his driver's license, just in case he got the urge to run again.

I didn't bother to look at him as he started to brag on himself.

"I was good, real good," he said.

I knew he was trying to get some sort of response out of me. I wasn't going to give him the satisfaction.

I just kept writing. I didn't even raise an eyebrow.

That frustrated him.

He obviously was some mama's little darling and had been told repeatedly he was oh so special and believed it.

Despite the fact I had not given him the slightest indication I was at all interested in hearing the story of his school boy triumphs, he just kept on talking.

Mr. Saturday Night and oh so special said, "I could outrun you—easy."

I could hear the sneer of youth in his voice.

Just 10 minutes earlier this kid had been hung up in barbed wire and crying for help like a six year old—practically in tears.

He might have been trussed up good enough to be stuck for days—maybe even lie and die there like some medieval prisoner in a dark dungeon.

But now that he's out of prison, he's back to acting like a Jack Russell terrier trying to pick a fight with a big dog three times his size and a heck of a lot wiser—me.

There's only him and me in the woods and still he has to show off.

"This kid sure is a piece of work," I'm thinking to myself—but I still don't let on that I even hear him.

I keep writing a good few minutes, and when I'm done, I hand him his citation for taking a deer out of season.

Then I fold up his billfold and hand it back to him without a word.

He's practically on his toes now, jabbering on about how he blew past the other kids on the 100 yard dash a few years ago, how he lettered in this and in that, what a big man on the OV campus he was.

It's making him crazy that I won't acknowledge him.

He might as well be a three year old screaming, "Pay attention to ME!"

When he finally stops bragging, I give him a

long look, lean back, stand tall and give him
my assessment.

I swing my flashlight beam back over to the
barbed wire fence.

His eyes follow the beam.

Threads from his torn up jeans, Tshirt and
underpants flutter in the night air above the
trampled brush and stomped blue clay.

I let the light linger there a few seconds.

His feet stop their little dog dance and his heels
settle on the ground.

I bring my light back to where the two of us are
standing.

He looks up into my face for an explanation,
silent at last.

"Track star, hunh?"

My voice drips equal parts boredom and
sarcasm as I look down at him—square in his
little black button terrier eyes.

He looks up into my face like a puppy who had just peed on the new carpet, but still wants his cookie.

"Well, you sure ain't worth a damn on the hurdles," I say plainly.

No malice, just a fact.

"Now, let's go get your buddy."

The kid never made another peep.

*"I grabbed the gun butt and pulled,
and got the muzzle up about an inch
before it slammed back down
into its sheepskin nest
like a terrified turtle
retreating into its shell."*

HARD TIMES

CONTRIBUTED BY HOWARD BROWN

Some readers may remember me as "Brownie"—the warden who gave a young fellow named Eric Nuse a dynamite idea on dead moose removal. The story is "Moose Vesuvius," in Volume 1 of this series.

I've been asked to share a few of my own adventures, and so here I am.

I began my career as a game warden in Maine in 1946. Later, I became a federal warden. Many a day was spent working here in Vermont.

But when I was a young man and just starting out, I was deep in the Maine woods.

It was just after World War II and I was totally on my own.

The challenge for wardens then, as it is now, was stopping the night hunters who wanted easy venison.

There were no cell phones, no radio communication from a warden to a supervisor. Many homes still didn't have any phone at all and those that did shared a party line.

You picked up the receiver and spoke to the operator—a real live person and probably someone you knew—and asked her to connect you with a neighbor down the road.

As for law enforcement, it was a bit of the Wild West—except there was a lot more moss and I wasn't on a horse.

The terrain was fir trees and bogs and ponds. I drove whatever used car I could afford to keep going.

Wardens were given the uniform, but we had to buy our own gun and use our own car.

Honestly, the years immediately following World War II were real tough.

Maine has never been a state where the workingman could make a lot of money. But when the war was on, men—and women—found jobs in steel mills and shipyards.

When the war ended, all those good jobs disappeared.

Add to that all the veterans returning home from the war, swelling the numbers of men looking for work.

Thousands of people were without jobs and struggling. Sometimes food was hard to come by and that made the temptation to take wildlife out of season all the more tempting.

My territory was 300 square miles and you have to remember driving the Maine roads back then was not anywhere close to being as easy as it is today.

I was based out of Lincoln—which is where the courthouse was at the time. That made it convenient when I had to appear and defend the cases I put before the judge.

Millinocket is about 23 miles north of Lincoln and Bangor about 50 miles off to the East.

There were no interstate highways and in fact most roads were still dirt—and that dirt turned to mud in the spring and fall, and deep snow in winter.

Cars were heavier, slower, lower and a lot less reliable. Power steering was a luxury.

Four wheel drive didn't exist in cars except for

maybe some surplus military Jeep a guy might get ahold of.

Motors and tires blew often and vehicles needed to be serviced every few thousand miles.

Hand chokes and carburetors were a nuisance and every car or truck had its own peculiar way of starting.

All this just to say it was a very different time. You had to be self reliant and frugal to get by.

When I started as a warden, the pay was $3,000 a year and the hours were 24 hours a day, seven days a week.

You got one week off a year for vacation. And that week off?

Wardens were strongly encouraged to take just one day off at a time.

Oh, and another thing: We were told we should be out in the woods at least five nights a week looking for poachers.

I was a young fellow with a new wife and baby. We didn't have much of anything—particularly money.

I didn't even own a side arm.

When they told me I had the job, I had to borrow a .22 caliber pistol from a family friend.

He didn't have a holster for the gun and I needed one so I could hang it from my belt on patrol.

I didn't want to spend the money to buy one.

I had saved some leather scraps. So I made a pattern of the revolver and cut it carefully, used an awl and some Aunt Lydia's thread and stitched together my own holster to save money.

To cushion the revolver and keep it from slapping around, I lined my holster with sheepskin and cut two slots up in the top of the leather so the holster would fit on my belt.

My holster didn't look bad if you didn't get too close. I was pretty pleased with my handiwork the first couple days on the job.

But I soon ran into trouble.

I just couldn't get the pistol to sit tight and stay put.

My .22 was forever spilling out of the sheep hide. It would slide across the car seat onto the floorboards when I grabbed for it in the dark on a late night stake out.

Poachers would get a lead on me as I fumbled around the car looking for the pistol in the dark.

A couple times I had to jump out and chase after guys shining my flashlight and shouting and just hope they didn't notice I didn't have a gun.

I was afraid my luck might run out. If they ever took a harder look at me, I knew I'd be in real trouble.

And worse, if they ever shot at me, I'd have no way of defending myself.

The badge alone was not going to impress some of these fellows.

I had to deal with some veterans who were what we called shell shocked. They call it post traumatic stress disorder today.

What it meant for me as a warden trying to enforce the law was some of these fellows were severely depressed, angry and had a hair trigger temper. They were out of work and out of sorts. Desperate.

They knew how to handle a gun and were trained to kill through their military service.

I knew if I wasn't careful with my words and actions a guy might just snap and kill me from all that pent up emotion that had nothing to do with me.

So, while I was strict and applied the law, I always made a point of treating everyone with respect.

I think that helped me over the years and may even have saved my life once or twice.

But in my first few months on the job, my biggest problem was my holster.

I tried all kinds of adjustments to keep my weapon from going AWOL.

It was one thing to have the .22 squirt out on to the front seat when I reached for it. It was a whole other mess to be running through the woods chasing after some fellows in the dark and have it bounce out.

I needed to tie the gun in the holster, but I knew

I couldn't be fiddling with trying to untie a string running through the woods.

What to do?

I hit upon the idea of making a rubber band.

I had a blown out bicycle inner tube in the garage—you remember we didn't throw anything away, right?

I cut a wide piece from the blown tube, and stretched and stitched it to my holster to keep my pistol secure inside.

Then I tipped the holster this way and that way and even upside down. The pistol sat right. It didn't budge.

I was pleased with myself and my frugal solution.

But a couple nights later, I found I'd done too good a job and about got myself killed.

It was August, with a Friday the 13th on the calendar. When I saw that I thought, "Well, that's supposed to be an unlucky day for somebody. I hope tonight it isn't me."

But I kissed my wife and baby and headed out for my nightly patrol.

I had a pretty good idea of what roads and fields the poachers in the area frequented. I drove to one of their favorites.

It was a narrow field with trees on both sides and an old hay road running through the middle of it about a mile from the main farm.

Poachers could drive in off the quiet dirt road, take a deer and then without even turning around, drive right out—like a rabbit den with a front and back door. It was a poacher's perfect set up and well known to them.

I rolled in around 11 pm with my headlights off and shone my flashlight just long enough to spot a doe out feeding.

Then I backed my old Pontiac off into a little cover of trees near one end of the looping farm road and just sat and waited—my holstered pistol on my hip this night.

I hadn't been there much more than an hour before I saw headlights come crawling up the road and someone shining a light out the passenger window.

I followed the beam as it passed over the field. No deer eyes reflected back. The doe I'd spotted earlier had moved on.

I stayed put and let the car roll on by me. I'd already learned that jackers often circle back.

Sure enough, about an hour later, here comes a car again.

And from the set of the headlights and sound of the engine, it seemed to me to be the same vehicle that was shining for deer earlier.

I saw someone shine a beam across the field. The light settled on a deer feeding not far from me—as easy shot.

Their car—it looked to be a Hudson Coupe— came to a quick stop and the brake lights shone red and full.

There was a bit of a moon out, not a lot, but enough so that I could see a rifle barrel poking out through the passenger side window.

The driver was getting himself set up to take a shot at the deer while the passenger blinded the animal with his flashlight.

My left hand went for the door latch and my right hand went for my gun.

I jumped out of my car and ran towards the back of the night hunters' Hudson and shouted out, "Warden! Stop right there!" and reached for my pistol.

I grabbed the gun butt and pulled, and got the muzzle up about an inch before it slammed back down into its sheepskin nest like a terrified turtle retreating into its shell.

What??

I am running right at their car.

They have a loaded rifle and who knows what other guns in there.

This is not a good time for me to be unarmed.

My eyes are focused on the rifle barrel sticking out the passenger window towards the deer.
If that barrel swings towards me, I'm in big trouble.

I tug again on the gun handle, harder this time.

I get the revolver up about two inches out of the

holster before it slams back into the sheepskin again.

It still isn't registering in my brain what the problem is with my gun.

I am focused on catching these guys and hoping to not get killed doing it.

Beads of nervous sweat are breaking out on my forehead even though frost is settling on the dead grass.

I start yelling at the poachers, "STOP! HALT!" not wanting them to shoot the deer or me.

The driver and the fellow with the rifle turn to look at me.

It's dark, but they can see something's not right with the way I am coming at them.

I'm running at them like the Hunchback of Notre Dame, my shoulder bent and all doubled over like I'm in pain.

They watch me grab at my hip, yank, jerk back, lean in again and pull.

My face is all twisted up in frustration.

The poachers just sit there—stunned.

I think maybe they couldn't believe their eyes —which helped me.

I get within 20 feet of the driver's door when the brake lights stop shining, the Hudson's engine roars to life and they try to race out of there.

"Stop or I will shoot your tires out," I shout.

Well, that was my plan anyway, but unless I can get my pistol out of my holster it was an empty threat.

It was if someone had buried a giant magnet in the fleece.

I'm practically on top of their car now with no weapon drawn to defend myself.

I have no idea how many fellows are in that big coupe or how desperate they are.

If one of them decides to point a rifle or pistol at me and pull the trigger, I could be all done.

I stop dead in my tracks, fall forward a step, reach across with my left hand and grab the pistol butt with both fists.

I pull as hard as I can.

My teeth are clenched tight and cold sweat is running into my eyes. I know I could be shot dead any second.

I hear the sound of rubber stretching and my brain finally kicks in as to what my problem is.

I'd forgotten all about my inner tube fix, and tonight it is doing too good a job.

I'm close enough to the poachers' car to touch the back fender.

I groan and yank with all my strength on the pistol grip with both hands.

There's a sharp snap and a crack across the back of my hand like my second grade teacher has taken a world class batter's swing and laid a ruler across my hand.

If I hadn't had a white knuckled death grip on the pistol's butt I would have dropped the gun.

I hear the tires spin and see tall grass spitting out from under the rear wheels of the Hudson.

Lucky for me this fellow had near bald tires.

With the field grass dampened by the night dew settling on the land, flooring the accelerator was the worst thing he could do—it made the tires spin as if he was on ice.

The car was laying a burn track in the grass.

With my revolver finally held tight in my right hand, I raised my weapon, lunged forward and stuck the barrel inside the window—aiming it right at the driver's head.

The rear tires were spinning so fast thick clouds of black smoke rose up from beneath the car.

These boys were about to set the field on fire.

I pushed the muzzle into the driver's ear and said, "Stop right here!"

I guess cold steel was what he needed to convince him to take his foot off the accelerator.

The engine powered down and the tires stopped spinning and smoking.

It was tough to see in the dark and the fog of burning rubber didn't help either.

But I thought I saw something go flying out of

the passenger window as the engine whine died down. I made a mental note to go look for more evidence later.

I had my flashlight with me and I shined it bright inside the car now, hoping to get whoever was in the vehicle to freeze like the deer they had been hunting a few minutes ago.

At last I could count heads. There was just the two of them, the driver and a shooter.

The fellow in the passenger seat was scrambling to toss a coat over the muzzle of a rifle barrel in the back seat of the coupe.

Too late.

"All right, Men. Hands up, please," I said.

They did as they were told and when I was comfortable there was no fight in them, I reached inside the car and lifted up the coat.

Beneath it was a near new Winchester lever action 30-30 caliber rifle—a nice lightweight weapon popular at the time with deer hunters.

I lifted the rifle out of their vehicle and found it

chocked full of bullets with one in the chamber
ready to fire.

I started adding up the infractions in my
head—I had plenty to charge them with
all right.

I ordered them to step out of the car and while
their hands were in the air, I asked, "Where's
your flashlight?"

The driver and passenger both feigned a look of
innocence. Finally the driver answered. With
a pouty lower lip and eyebrows raised, he said,
"We don't have one."

I've never liked liars.

I sighed and said, "Well I saw you two shining
a light about five minutes ago at a deer in this
field."

I waited for them to confess to having a light.

No response.

"Well, how about I just search your car then?"
I said.

I emptied their pockets and then let them put

their hands on the hood while I searched inside for their flashlight.

I found a good selection of beer bottles, fishing lures and even some swim trunks but the light had indeed gone missing.

I realized the shadowy something I thought I saw flying out of the car some minutes ago was probably it.

This was no time to be looking for it.

Too dark and I didn't want to keep the pair of them around. They might just decide to try and fight me.

We'd let the judge decide.

As I wrote down their names, addresses and phone numbers, I noted they both had the same last name.

"You two brothers?" I asked.

They nodded in the affirmative.

"Well, I'm going to write both of you up for shining a light, attempting to take deer at night, carrying a loaded rifle in this vehicle and attempting to elude an officer of the law."

They looked stunned to hear all the charges.

"I will see you at the court house in Lincoln at 10 am sharp Monday and we'll see what the judge says about your behavior tonight. I strongly urge you to appear or there will be arrest warrants issued for each of you."

The two of them looked at each other, then back at me, nodded and looked then down at their shoes.

Then the taller of the two piped up.

"What about my rifle?" he asked.

Well, now I knew which one owned the Winchester too.

"Your Winchester is coming with me as evidence," I said. "It will be at the court Monday along with me. The judge will decide if you get it back."

"You can go," I added. "If you don't floor the gas you should be able to drive out of here."

After a little rocking, they got the Hudson out of the dip they'd dug.

I waited until the pair drove off, then I unloaded

the rifle, pocketed the shells and went back to my car.

In about four hours the sun would be peeking over the horizon.

I decided to stay put and sleep in my car. Come daylight, I'd walk the tree line and look for the flashlight and any other evidence they might have tossed out while trying to get away.

I loosened my belt and took off my holster and laid it and my pistol beside me on the seat. My right hand was throbbing and I could see there was some swelling but it didn't stop me from falling asleep.

I awoke with the sun shining bright in my eyes through the windshield. I checked my watch —it was a little before 7 am.

Once I could see clearly, I took a good look at the back of my right hand.

There was a yellow welt about two inches wide and an inch high stretching from my thumb to my pinky. I'd end up black and blue there for a few days.

Oh well.

I shook my head and chuckled a bit at myself.
Live and learn.

I walked over to where the night hunters had
torn up the field, and then walked back slowly
looking in the bushes and along the tree line
for evidence.

It didn't take long for me to find what they had
thrown out the passenger window.

It was a nice police quality flashlight, powered
by four big D cell batteries.

It was a lot more rugged than the one I was
using. I picked it up and grinned. I knew it
would help me make my case on Monday.

I looked around briefly for more evidence,
found none and headed on home.

Once inside the door, I made myself a
couple sandwiches and then filled a basin
with hot water and a good big dose of
Epsom salts.

I set the basin on the right side of the kitchen
table and the sandwiches on my left. Then I
pulled up a chair and sat down to relax.

As I ate with my left hand, I soaked my right

hand in the hot bath and let the salts do their
job.

Sitting there gave me time to ponder
the meaning of "penny wise and pound foolish."

I decided to find and buy a holster to fit that
.22 with my next paycheck, before I got myself
killed.

When I arrived at the courthouse Monday
morning, the brothers were there waiting.

Back then, it was pretty rare for someone to
skip a court date. Defendants all wanted their
guns back and losing the right to hunt and the
possibility of arrest for failing to appear were
all good reasons for defendants to just face
the music.

The judge at the Lincoln court was a buddy
of mine. He was a little fellow with a taste
for whisky after a day of partridge hunting or
fishing.

Maybe today our friendship would be called
undue influence or something.

But I don't believe our occasional hunting and
fishing forays together ever interfered with our
work.

I had a feeling these two fellows had been counting on me not finding their flashlight and would use that as part of their defense.

But if that was their plan, all hope was lost when they saw me sitting across the aisle pretty as you please with that shiny silver torch laid right next to their Winchester .30-.30.

Their faces fell like two hungry boys who'd just been told they were too late for chocolate cake.

With the evidence in hand, I told the judge my story—leaving out the part about my holster problem. It didn't take long for the judge to find them guilty.

The fine was $15 plus $4.80 court costs each and they lost their right to hunt for a year.

Because that was a pretty hefty fine back in the 1940s, they asked to pay the fine in installments to the court and the judge agreed.

When it was over, the judge took a brief recess.

As soon as the judge was gone, the pair of them walked up to me and asked for their Winchester rifle and their flashlight back.

"You can have your rifle—after you've paid your fine in full," I said.

"But you fellows told me you didn't have a flashlight with you, right?

So the one I found there the next morning, this one right here, it can't be yours."

They looked at me, then one another in silence.

"Because if that was the case, well, then you would have just lied to me and to the judge and that's another offense—perjury."

They stood there in silence as their brains kicked into high gear trying to figure out how they could get that expensive flashlight back without admitting they'd lied.

I gave them a few minutes alone. They whispered to one another off in a corner while I waited.

It took a while but eventually they figured out there was no way for them to lay claim to the torch without causing themselves more legal trouble.

The two of them walked back over to where I was standing with all the enthusiasm of men headed to the electric chair.

The older brother was the only one who spoke. "We're done," he said.

I wanted them to know it was the lying that had cost them.

I held up that big barreled shiny flashlight just inches from under their noses, smiled and said, "She sure is a beauty. And since no one's laid claim to it, she's the property of the state of Maine now, Gentlemen."

"Yessiree. This flashlight is gonna come in mighty handy helping me catch poachers."

And it did.

"I stepped into the cabin and saw what looked like a simple wooden kitchen table and off to the right behind a half wall was a big pot boiling on the kitchen stove."

SEARCH WARRANT

CONTRIBUTED BY HOWARD BROWN

I had a tip there was some poaching going on over on one of the many ponds near Lincoln, Maine where I was based out of in the 1940s and 50s.

Lincoln boasts around a dozen ponds and lakes close by. Cold Stream, Egg, Folsom, Crooked, Stump, Caribou and Mattanawcook are some of their names. I'll just let you guess as to which one it was.

I was told this poaching had grown into quite a sizeable operation—a kind of no tell butcher shop with suppliers and buyers and venison being bought and bartered among family and friends.

I made my case to the judge—my occasional hunting and fishing buddy—and based on the evidence I presented to him, he had no qualms signing a search warrant for me.

I thanked him, tucked the paperwork inside
my coat, jumped behind the wheel of my car
and went on down the road and around the
mountain to the address I had been given.

Problem was, it was darn difficult to figure out
which camp or shack was the right one.

There were dozens of them—some on the
lakefront and others on the hill across the road.
Stacked right close together too.

It was like driving through a Monopoly board
game and looking at all those little house pieces
tossed willy nilly across it.

There were very few names on any of the camps.

I didn't want to ride up and down the road slow
just looking.

This was a tight knit community. The folks there
knew each other's vehicles. I would make them
pretty suspicious and worse, give the guilty ones
time to start hiding evidence.

I finally just gave it my best guess. I stopped at a
camp that fit the description I had been given.

There was a rickety shack out front and a new

something or other—maybe a bigger camp or maybe a garage—under construction out back. I walked up, knocked on the door to the screened porch and a woman in an apron, maybe 40 years old, answered and I introduced myself.

"Good morning, Ma'am. I'm Howard Brown the fish and game warden for the area and I have a search warrant authorizing me to search property here."

I started to reach into my breast pocket to pull the paperwork out and show her. The lady's eyes got huge and she kinda stumbled back a step.

She shook her head no, like she didn't want to see it.

Some people do find that sort of thing really intimidating.

So, I just tucked the paper back into my pocket and waited.

She didn't say a word. She just opened the screen door wide for me, stepped back another big step and let me in.

The first thing I noticed was the smell.

I stepped into the cabin and saw what looked like a simple wooden kitchen table and off to the right behind a half wall was a big pot boiling on the kitchen stove.

The steam wafting up from that pot contained the unmistakable smell of onions, carrots and venison.

I stepped up to take a look at the stove, looked back at her wringing her hands on her apron and saw a look on her face I can only characterize as guilty as sin and resigned to her fate.

"Why don't you just take a seat at the kitchen table here while I look around? Would that be all right?" I asked.

She just nodded and scuttled across the worn linoleum, pulled out a beat up wooden chair and sat down in it as fast as if we were playing Musical Chairs.

Her back went rigid, her head was bowed, her hands folded neatly in front of her as if in prayer.

"Anyone else home?" I asked her.

She kept her eyes on the worn tabletop and

shook her head slowly side to side in a big "No."

"Okay. You stay right there, please," I said. "I'm going to look around here a little."

It was a small camp. I walked into two different bedrooms and found a rifle propped against the wall in each of them.

It didn't take me long to find parts of five different deer hanging or partially cut up— most of them hidden in the new addition being constructed out back.

Maybe they intended for that new building to serve as their store.

I made my observations, took some notes, then came back into the main camp and found the woman still looking stunned and sitting still as a statue at the kitchen table.

The venison stew was still simmering. It smelled real good.

I walked over and lifted the pot lid and saw carrots and potatoes and onions all tucked in there along with big chunks of venison.

I turned to the woman and said, "Well, Ma'am, I found a lot of evidence here of deer jacking.

I am going to have to write your husband, Bill, up for taking deer out of season."

Her head rocked back and she looked at me hard. For the first time since I met her 20 minutes earlier, she spoke.

"Did you say 'Bill'?" she asked.

"Yes," I said. "Your husband is Bill Peterson, right?" I asked.

"No," she said. "My husband is Jacques Reynaud. The Peterson's camp is around the bend."

"Oh," I said.

It began to dawn on me that what I had here was an enclave of deer jackers and that the search warrant in my jacket pocket might be put to broader and better use.

I knew that I had to move fast if I was going to find Mr. Peterson's camp.

"Well, thank you." I said. "I'll be back later."

I knew if I drove back to the courthouse and asked the judge for another search warrant, word would get out fast about me snooping around—even without the benefit of telephones.

My suspects would all disappear or at least most of the evidence would.

So, I made a decision to just keep going.

I knocked on some more doors in the neighborhood and asked to look around.

Each time I tapped the search warrant in my shirt pocket and told whoever answered the door the truth: "I've got a search warrant here," followed by, "I'd like to come inside and look around if I may."

No one asked to look at the warrant, let alone take the time to read it and see if it was written for their camp.

I could tell just by the reaction on some folk's faces when they answered their door that something hinky was going on inside.

But it didn't take long for news of my visit to the neighborhood to spread like wildfire.

Curtains got pulled and kids were called indoors and then dispatched on foot and on bicycles to other camps. I knew they were warning other camp owners about me being in the neighborhood.

Some folks just locked their doors, jumped in their cars and drove away fast.

My informant lived here too. To cover his tracks and mine, I paid a visit to his home using that search warrant too.

By the time I was done, I had looked inside a couple dozen camps.

I found evidence that at least eight camp owners were somehow connected to taking, selling or buying deer.

For those who weren't cooking venison on the stove to give themselves away, there were others with ice boxes stuffed with venison roasts, steaks or hamburg inside.

Others had fresh deer hides lying out back or stuffed into a garbage can.

With this information in hand, I went back to the judge, told him what I knew and asked for search warrants for each of the camps where I had seen evidence of poaching or possession of deer meat.

Then I went back and did a more thorough search.

Before I was done, I had cited more than a dozen of the locals along that winding shore road into court.

I call that one my All Purpose Search.

"*This portrait of matronly modesty tore into me,
my parentage, my heritage and my profession
as if I was Attila the Hun.*"

HAVIN' HER SAY

CONTRIBUTED BY HOWARD BROWN

I was quietly checking fishing licenses out on Cold Stream Pond near Enfield, Maine by canoe.

There was no big warden sign on the canoe or me. I just sort of paddled about quietly and spoke to folks, asked for the licenses if I saw them fishing.

I'd been keeping an eye on one 16 foot rowboat for a few hours as I made my way around the pond. There were three people in this boat—two men and one woman.

The fellows were fishing all the time I was out on the water, but the woman just sat there most of the day. But by the time I paddled up to their boat in the late afternoon, I saw she had finally wetted a line too.

I introduced myself and asked all three for their fishing licenses. The men quickly produced theirs, but the woman said she had left hers back at the camp.

"Well, you are supposed to carry it with you,"
I said. "But let's go on over to your camp and I
will let you find it and show it to me."

She looked a little put out, but the fellows reeled
in their lines and pulled up the anchor. The
fellow in the stern began rowing the boat over
to their camp. I followed.

She got out of the rowboat at the dock and I
pushed the nose of my canoe into the shore,
hopped onto a rock and proceeded to follow her
up the wooden steps into the camp.

"We don't fish much," she said as she pulled
open the screen porch door.

I stepped inside and the walls were lined with
more than two dozen fishing poles of various
sizes and vintages. Beneath and surrounding
them were six or more tackle boxes filled with
spinner baits, spoons and sinkers.

I didn't say a word.

I let her dig around the camp for ten minutes
or more, tossing papers and moving towels and
even rummaging through the medicine cabinet—
until she came to accept the fact I was not going
to tell her to just forget about it and thanks for
trying.

Exasperated and angry, she finally admitted to me she didn't buy a fishing license that year.

I just nodded and wrote her a citation and told her to appear in the Lincoln courthouse the next day at 10 am before the judge.

She did one of those, "Well, I never!" sort of insulted women stomps when she took the ticket from my hand.

I just said, "Good day, Ma'am" and left.

She was at the courthouse bright and early the next morning. She was all dressed up in Sunday go to meetin' finest with white gloves and lady shoes with smart heels on 'em and even a purse to match.

Her eyes flashed fire when she looked in my direction and her shoulders swayed as if she was shaking off a pesky fly.

It was clear to me she was getting herself set to unleash a storm like a home run hitter in the bottom of the ninth in the final game of the World Series.

Her lips were pressed so tight together they looked white even with a big smear of red lipstick.

She looked like she was ready to bite the heads off a pound or two of six penny nails.

All around us are the usual deer jackers I'd nabbed too—most of whom were unshaven and wearing dirty jackets and scuffed boots and frayed trousers.

This lady looked outta place among them. But just like those guys, she had been caught breaking the law and I had a duty to treat her the same as all the others, which I did.

When it was her turn before the judge, she said again that she hardly ever wetted the line on a fishing pole so, no, she had not bought a license yet this year.

When the judge asked her, she admitted she was drowning a worm when I floated on up to their boat. But she claimed she hadn't so much as even had a perch tickle its tail.

The judge listened to her excuses with what looked like real compassion.

Then he shook his head in sorrow, hemmed and hawed a bit and explained that as sympathetic as he was to her testimony, under the law, his hands were tied, he had no choice but to find her guilty.

He fined the lady $15 and $4.70 in court costs, just like all the others found guilty that day.

When she heard the verdict, her head snapped back as if a glass of cold water had been tossed into her face.

It was clear she didn't expect to be treated the same as all the ruffians men surrounding her.

"Well!" she said in exasperation. "Can I say something now, your Honor? Is it my turn?"

As a rule, the judge didn't allow post conviction comments from defendants for fish and game violations.

But I guess seeing as how she was dressed for church—or maybe loaded for bear—for whatever reason, he said, "You may address the court."

That was all the encouragement this woman needed.

The lady cast a glance at me off to her right that was as dark as any ever cast upon me.

A starved panther about to pounce onto the back of a skipperjack and sink its fangs into its jugular could not have looked more purely predatory.

This portrait of matronly modesty tore
into me, my parentage, my heritage and my
profession as if I was Attila the Hun.

She said her family had owned that camp for
more than 50 years, her clan lived and worked
in the area since the state was first settled and
if all that settling and taxpaying and working
and birthing and dying didn't allow one little
lady like herself to dip a line in the water
for an hour or so once a decade without an
overzealous game warden like that buffoon over
there swooping down on her—that would be
ME she was talking about—well, what was the
world coming to?

She ranted and raved a good five minutes.

All the time the judge is looking at her nodding
and his hands are folded and his fingers knitted
together.

"What the heck is this?" I'm thinking to myself.

If any fellow tried to go on like this the judge
would bang his gavel and rule them out of
order.

When the lady had blown off enough steam to
heat the courthouse for an entire Maine winter,
the judge said, "The court will take your opinion

under advisement," followed by, "We will now take an hour recess. The court is adjourned until 1 pm."

The court clerk said, "All rise," and we did. And as soon as the lady in white saw the black robed judge shut the door to the judge's chamber, she gave one last dagger eyed glare at me.

Then she stomped out of the courthouse with her head held high, as triumphantly as Joan of Arc.

The men who were waiting their turn before the judge looked at her in awe.

I bit my lip and hoped this lady wasn't the start of a trend.

Guilty is guilty in my book.

I waited until the courtroom cleared out and then made my way to the judge's chamber.

I knocked and heard a "Come on in, Brownie," from inside.

I entered to find the black robed bait fisherman smiling like a Cheshire cat. He'd already lit up a cigar and was sitting behind his desk enjoying it.

"What was that, Judge?" I asked. I wasn't very happy.

"That, Warden Brown, was the 30 cents change due the lady from her $15 fine and court costs."

"Well, Your Honor, I think you not only gave the lady her change but a nice fat tip too," I said.

The judge chuckled, leaned back in his black leather chair and pulled two shot glasses from a desk drawer. Then he reached behind him and poured us each a shot from a whisky bottle he kept on a shelf above a long row of law books.

He took a deep breath and then explained his thinking.

"There's a difference between your job and mine, Brownie, and I don't just mean the black robe here. You're a state employee. Unless you do something awful bad, you've probably got a job for life."

I listened and nodded.

"But me? I'm a political appointee.
The Legislature decides whether or not
I get to keep my job. And if enough people complain, I might not have it the next time they vote."

I take a sip of the whisky and remain silent.

"That lady you brought in here? She's from a big political family. Got a brother who's a senator. Got a former governor in the family back there somewhere in her pedigree too, I believe."

I say, "I don't think that should make any difference," before the judge cuts me off.

"Oh, don't get me wrong, Brownie. She's guilty all right. You were right to bring her in. But sometimes you gotta let certain folks blow off a little steam—give the people their money's worth, I guess, you'd call it."

The judge saw the scowl on my face and laughed.

"It's politics, Brownie," the judge laughed.

"Now drink up and then let's both of us get back to work."

"We bury the bird and we don't go to jail?"
the short one asked."

Skinny Goose

Contributed by Howard Brown

It was the first day of goose hunting in Vermont, early September, and for me that meant heading out into the dark around 4 am to patrol the shores of southern Lake Champlain.

It's important to ride herd on the crowd of eager waterfowl hunters, especially on opening day. Too often there's one or more guys with itchy trigger fingers that just can't seem to wait for the official start of the season.

Most every year I would have to cite someone for shooting too early. Granted, it's a little tricky. The regulations call for guns to be silent until one half hour before sunrise.

As sunrise changes daily, its important to keep a copy of the regulations in your pocket, have a good watch and synchronize it with a reliable source.

And regardless of the time, you need to know what you're shooting at.

The rules aren't there just to give the ducks
and geese a little more shuteye. They're there
to protect hunters from shooting each other.
An early shooter causes not just frustration for
the other fellows, but can put them in
real danger.

Setting up before dawn in the reeds, wearing
camo, floating decoys and then calling in ducks
and geese—waterfowl hunting is a war waiting to
happen if someone cuts loose with buckshot in
the dark.

And of course, if one fellow shoots early at a
bird, then the birds spook and fly up and that
sparks another fellow to shoot and so on down
the lake or pond until it sounds like a war zone.

This before dawn visit, I was headed down to
the Dead Creek area of Addison County off
Route 17 and the big watershed adjoining Lake
Champlain. It's a popular and easily accessible
area for duck and goose hunters.

Thanks to a lot of restoration work, thousands
of geese now fly up and down the lake each
year in their migrations from Canada to the
Chesapeake Bay.

In earlier years, I spent many summers
banding ducks and geese in Canada—from

Saskatchewan to Labrador—to aid in the effort of restoring these species.

This day, I decided to play it low key. The hunters didn't want to spook the geese and I didn't want to spook the hunters.

I drove down near Panton in my civvies and an unmarked car. My badge and revolver were hidden under a coat that ended just above my knees.

The ducks and geese in this area didn't really have a fair fight in some ways. If a hunter was lucky, no duck blind or hike into any wet land was even necessary.

You could pull off onto the grassy strip between the roadway and the shoreline, walk a few feet past the trees and peer over the cattails and fire at will at whatever was sitting there or flying above on a good day. No need to even get your feet wet if you were real lucky.

All was quiet on the west shore road as well it should be. It was a good 10 minutes before the start of the hunt.

But as I rounded a bend and wended my way up the east shore, I saw the night sky was lifting its veil and there was just the slightest hint of the dawn to come in the East.

My headlights shone on the outline of a truck maybe 100 yards ahead of me and parked off on the grass to my left. There appeared to be at least two figures beside it carrying shotguns.

I slowed my car and a split second later saw a flash of fire from the muzzle of one shotgun followed a split second later by a second blast less than 10 feet away.

I pulled my car over, killed the engine and my headlights. I turned on the dome light in my car and looked at my wristwatch. It was a good five minutes before the start of the season. Time for a little chat with these fellows.

I tucked my citation book inside my jacket, grabbed my flashlight and climbed out of the car.

I didn't rush in on them. I just walked on up and asked, "Did you get him?"

"Two geese flew. I think we got one at least," the shorter of the two fellows said with a big grin of accomplishment on his face.

"Not bad for the first couple minutes of the season," the taller fellow piped up, hoisting his shotgun to his shoulder and standing up tall.

The shorter guy broke open his still smoking

double barreled shotgun and popped the empty cartridges into his coat pocket.

"Well, what are you waitin' for?" I asked. "You gonna go get 'em?"

"Well, it's still a little dark to walk out there and find the bird. I'm not quite sure where it landed, either," the taller fellow admitted, his grin fading.

"I've got a flashlight here. I think I can find him for you," I said. "You just wait right here."

I pulled my flashlight out from under my coat.

They smiled big at me and said, "Thanks!"

I headed out into the brush and hit the marsh edge just 30 feet inside the trees. The rising sun helped me some.

Within three minutes I had the dead bird in hand and was headed back to them.

They were in for a little surprise.

I broke back through the brushy roadside maybe 10 feet from where I ducked in five minutes earlier.

They were beaming like a couple boys who just

taught the family dog to retrieve a stick.

A stranger doing all the work for them? How lucky could they be?

But what this retriever was carrying was not what they expected.

"Let me see it!," the short guy said. "Wow, it's big!" the other guy said as I came out of the shadows.

But the expressions on their faces changed the closer I got to them.

They couldn't have been more surprised if I had brought them a six foot alligator.

"Oh my God! What is THAT?," screeched the taller of the two and he jumped back a good yard in fright, slamming the small of his back so hard into the truck grill he about fell down.

His hunting buddy gasped, uttered an "Awk!" like he was choking on a minnow, leaped back about three feet and almost dropped his shotgun.

"Oh crap!" he said.

I was holding not a goose, but a heron—dangling

a good five feet from beak to big toe—colored like the shingles on a 100 year old weathered Cape Cod beach house, and only about as big around as my fist.

Basically, a walking stick with wings.

There wasn't any more meat on this heron then there would be on a gray squirrel.

And unlike squirrels, Great Blue Herons are a federally protected species.

These boys were in some trouble.

Shooting early, shooting a protected species and shooting from the road to be exact.

But I just stood there letting it all sink in.
"We shot THAT?" the tall fellow asked unable to believe his eyes.

"That ain't no goose—is it, Joe?" he said turning to his buddy.

The shorter fellow set his shotgun against a tree and stepped in for a closer look. I held the bird out in front of me and let him inspect it.

He stayed about three feet away as if the long bill on that bird might stab him.

Then he bent down a bit, grimaced like he was looking at bear scat in the woods, and looked the lifeless bird up and down.

He stepped back, stood up straight and gave his opinion.

"I think it's some kind of crane, Billy," he said.

Then the pair looked up at me—seeking an opinion from their retriever.

"That would be correct," I said. "A great blue heron to be exact."

All around the pond now the place erupted with shotgun blasts.

I looked at my watch. Yup, the other shooters knew how to tell time.

Joe and Bill stood staring at each other.

"What are we gonna do with this thing?" Bill said.

"I don't think we can eat it. Is it even legal?" I saw both of them were a good distance away from their weapons and before their discussion got too involved, I decided to give them the bad news.

"This heron is a federally protected bird," I said. "If you'd just waited until the start of the season this morning, you would have been able to tell a heron from a goose."

They looked at me like I was a talking dog and a smart aleck dog at that.

I saw both of them scowl like they didn't want to have to listen to a lecture from their retriever.

"Basically, you've done three things wrong this morning and it isn't even 6 am," I said, finally lowering the bird to the ground.

"Three?" the shorter fellow said looking up at me.

"Yes, three," I said.

"You shot before the start of the season—both of you. And you shot a protected species."

"Well, okay, that's two," the tall fellow reluctantly admitted.

"Three," I repeated. "I'm a game warden," and I unzipped my coat to show them the badge on my shirt.

The tall fellow—who had just gotten his feet

back underneath him and was trying to come up
with a plan as to what to do with this bird—went
stumbling backwards again into the truck grill.

He turned and pounded a fist on the hood.

"Dang it! I knew I shoulda just ignored you,
Joey, and stayed in bed this morning," he
moaned. "Look what you've gotten us into."

Joe started to defend himself.

"Me? You're the one that always said you
wanted to try goose hunting. I was doin' you
a favor!"

I could see these two coming to blows pretty fast
if I didn't stop it.

"Just settle down, Gentlemen, and show me
your licenses and tags, please," I said.

By now, there was a bit of a crowd around us
—as some of the other hunters had seen me
walking out with the great blue and wondered
who in their right mind would shoot one and
then be dumb enough to walk out onto the
roadway with it for all the world to see.

There were fellows on the other side of the road
watching this unfold.

I didn't mind at all as long as they didn't block the road.

Having an audience meant the word would spread quickly that hunters need to follow all the rules.

While Billy and Joe stood there in silence, I checked each man's license.

They were valid, along with their duck and goose tags.

I copied their information and seeing as how the tall skinny guy looked to me like he was about to throw up or just fall down dead he was so pale, I decided to cut them a bit of a break for their stupidity.

I also knew something they didn't that I wasn't about to share.

That heron wasn't in great shape—and I don't mean because he'd just been knocked out of the sky by a shotgun blast or two.

I saw his feathers were tattered and he was significantly lighter in weight than he should have been after a summer feeding on an all you can eat Dead Creek buffet of frogs and minnows.

I figured this bird didn't have a lot of flying and feeding left in him. I couldn't see him going into winter in this shape and making it to Spring.

And if I was to cite these two with killing the heron, and they fought the charge, I'd have to carry this big bird in my evidence bag today and then stuff him in a freezer until their case made it to the court docket.

Keeping him on ice when I had a whole season of gathering evidence ahead of me—well, I decided maybe I could just teach them a lesson.

"Well, gentlemen, I could throw the book at both of you here. Your killing this bird—a protected species—could get you six months in jail and a $10,000 fine."

That got their attention. I thought the tall fellow was going to faint. Before his knees buckled, I spoke again.

"But I have a proposition for you, if you're interested. I am going to cite each of you for discharging your firearms before the start of the season, but I'll waive the much bigger possible fine and jail time if you'll do the right thing by this bird you killed."

They stared at me like men adrift at sea looking

reaching out for a lifeline I was about to toss 'em.

"I've got a shovel in the back of my car. You take this heron you killed and bury it nice and deep over near the edge of the marsh.

You do that—and you study this bird real good as you bury him—and maybe you'll learn a little about the difference between a heron and a goose."

I saw them trying to comprehend my offer. "We bury the bird and we don't go to jail?" the short one asked.

"You're both going to get cited for shooting before the start of the season, but yes, I won't charge you with the bigger crime of killing this protected species if you do a good job burying the bird," I said.

The two of them came to life fast when the deal I was offering them sunk in.

The skinny fellow—now that I think about it he kinda was built like a crane himself—about fell to his knees with relief.

The shorter one, said, "Yessir, Warden. That works for me! Thank you! Thank you!" and a big grin came over his face too.

"Okay, come along with me to my car and I'll get that spade for you."

The shorter fellow was jigging from one foot to the other like a bird dog waiting for a command.

I walked over to my car with the two of them hot on my heels, popped the trunk and dug out my shovel.

"Dig the grave at least two feet deep—do a good job, nice and neat. I'll be down to check on you in a bit," I said.

The shorter one took the shovel and his taller buddy followed.

Of course, they had to take the bird with them too.

I stood there trying not to laugh as the shorter guy slowly bent down to pick up the heron's long floppy neck with his free hand. He spread his thumb and forefinger like he was a cherry picker loading a log bound for the Ticonderoga paper mill.

It was pretty clear he hadn't so much as ever petted a parakeet before. He made a face, then pinched his fingers together—just behind the bird's head.

He acted like he was picking up a rattlesnake—
a live one.

When he finally clamped ahold of the bird, he
nodded to his buddy who squatted down like he
was about to pick up a 50 pound bag of grain.

He picked up the heron's legs—one in each hand
like the bird was a human—and then he stood
up straight and nodded to the shovel man.

His jaw about hit his chest when he stood
up and realized the bird weighed about five
pounds.

"There's nothing to this bird but feathers," he
said in amazement.

With a good three feet of rail thin bird between
them, it's long slate colored wing tips dragging
a trail in the roadside dirt, they shuffled off into
the tree line, into the marsh grass and out of
my sight.

While the pair were off in the brush, I made
some notes, wrote down the truck license plate
number, unloaded their shotguns and stowed
them in the trunk of my car.

Then I headed on down to see how the grave
digging was going.

The Addison County clay was good and wet.
It might have been heavy to move, but it made
for a nice looking grave with clean edges. They'd
about finished up the hole by the time I go
there.

Maybe it's just human nature or how we are
raised. But when they laid the bird neatly in the
grave, the two of them stood there as if this bird
was someone they knew lying prostrate in the
hole.

There was an awkward silence.

Then the tall skinny fellow reached up and
removed his cap and the shorter guy leaning on
the shovel muttered, "I'm sorry, Bird."

Maybe they did this just for my benefit.
Maybe when they were five years old they'd
buried a pet goldfish in the back yard and were
remembering that experience. Or just maybe
they really meant it.

It was a nice touch anyhow.

I gave the bird a moment of silence and then
said, "Okay, go ahead and cover him up."

The shorter guy knelt down and began pushing
the wet earth and swamp grass over the top of

the remains with his hands. His taller buddy grabbed the shovel and began laying in clods of clay. When they were done, I stood back and took a good look.

It was a funny shaped grave. It looked like someone had gone to a lot of trouble to bury a five foot long, 12 inch wide plank or some crazy thing.

They both turned around and looked up at me waiting for the nod that they'd done the job right.

"Okay, fellows, that's good. Now, let's go on up to my car and I'll get you your paperwork," I said.

I took my shovel out of the little guy's hand and used it as a walking stick for the short trip back. I had them walk ahead of me.

Back at the truck, I handed them each a citation for shooting before the season started, told them they were done goose hunting for the year and that they needed to show up in court if they expected to ever see their shotguns again.

Then I gave them what I hoped was some advice they'd take to heart.

"You two don't seem like bad guys.

But you made some serious mistakes here today. I suggest you each get yourselves a wristwatch, read up on the wildlife regulations and carry it with you and study up on what ducks and geese look like on the wing too."

"Okay, Warden! Thank you," the shorter fellow said.

"Me too. Thank you, Warden. I sure will," the taller guy said and reached out to shake my hand.

They nodded and smiled and jumped into their truck as happy as if I'd been a passing tow truck driver who pulled their truck out of the ditch in a blizzard on Christmas eve and not charged 'em a dime.

I crossed the road ready to continue on my way. I had a canoe waiting for me in the reeds a mile or so up the road. I planned on visiting some duck blinds this morning.

I had just opened my driver's door when a young fellow in waders and carrying a shotgun over his shoulder strolled on up to me with a big grin on his face.

I'd noticed him earlier kinda hanging back and watching me deal with the heron shooters.

"I gotta tell you, Warden. I've been goose huntin' since I was a kid—started at age eight alongside my dad. I've seen maybe a dozen hunters get pinched by various wardens out here over the years."

He paused and looked at me with a twinkle in his eye.

"I sure never heard any of them thank a warden for giving them a ticket and taking their guns. You must be one of the good guys."

He walked off into the reeds chuckling to himself.

I smiled and headed on up the road.

That was one of the nicest compliments I ever got.

Acknowledgement

*This book could not have been completed
without the encouragement and skill
of the following individuals:*

*Jean Pamela Poland, Dorrice Griffith Hammer,
Sandy Brisson, W. Douglas Darby,
Paul A. Young, Stephen P. Frost,
Carrie Cook, Norma Montaigne,
Al and Karen Myers, Eric Nuse,
Ingrid Nuse and O.C.*

Thank you one and all.

Many thanks to the following wardens...

 **Eric Nuse** worked 32 years as a warden in Vermont. Eric resides in Johnson, VT.

 **Richard Hislop** worked 34 years as a Vermont warden. Richard resides in Fairfax, VT.

 **Stan Holmquist** worked 27 years as a Vermont warden. Stan resides in Rochester, VT.

...FOR ALLOWING ME TO SHARE THEIR STORIES

 **Denny Gaiotti** worked 29 years as a Vermont warden. Denny resides in Whiting, VT.

 **Howard Brown** worked 31 years as a state and federal warden. Howard resides in Swanton, VT.

Stories by Warden

Eric Nuse
4WD
Stowe Turkey

Richard Hislop
Tick Trail
Blame the Name?
That'll Teach Him
Dog Fish
Stay
Wrong is Right

Stan Holmquist
Fly Trap

Denny Gaiotti
Track Star

Howard Brown
Hard Times
Search Warrant
Havin' Her Say
Skinny Goose

WHO WE ARE

 **Megan Price,** the author, is a former award winning journalist and Vermont legislator who knows a good story when she hears one.

 **Norma Montaigne** is an accomplished illustrator and graphic artist. She lives and works in Pittsford, Vermont.

 **Carrie Cook** is an exceptional graphic designer and musician who lives in Cambridge, Vermont.

Want more great stories?
*Read **Volume One** yet?*
Here's what you're missing:

FISHIN' TACKLE

RACCOON RIOT

MONGO

CLYDE RIVER RACE

HUNT 'ER UP

FURRY FISH FINDER

BEAR? *WHERE???*

SQUISH IN THE NIGHT

THIN ICE

TOO LOOSE MOOSE

MOOSE VESUVIUS

GIMMEE THE GUN

COVER ME

Volume Three coming soon...

Visit us at www.VermontWild.com